Jasmin Dierkes
Professionalisation of Medical Education

Science Studies

Jasmin Dierkes, born in 1995, is a sociologist and STS scholar who works on medical sociology, the sociology of professions and educational studies, employing contemporary and historical analysis. From 2022 to 2025, she completed her doctorate at Bergische Universität Wuppertal. Her recent work on simulation-based training in medical education sheds light on the training techniques and technologies of a 'classical profession', representing subtle changes in professional teaching and working standards.

Jasmin Dierkes

Professionalisation of Medical Education

The Rise of Simulation-Based Training

[transcript]

This study is based on a dissertation that was defended in 2025 at the University of Wuppertal as part of the GRK 2696, which was funded by the German Research Foundation (DFG).

Bibliographic information published by the Deutsche Nationalbibliothek
The Deutsche Nationalbibliothek lists this publication in the Deutsche Nationalbibliografie; detailed bibliographic data are available online at https://dnb.dnb.de

Hermannstraße 26 | D-33602 Bielefeld | live@transcript-verlag.de

Cover design: Maria Arndt
Cover illustration: @JamesReim/pixabay.de
Proofreading: Sean O' Dubhghaill
Printing: Elanders Waiblingen GmbH, Waiblingen
https://doi.org/10.14361/9783839478349
Print-ISBN: 978-3-8376-8231-1 | PDF-ISBN: 978-3-8394-7834-9
ISSN of series: 2703-1543 | eISSN of series: 2703-1551

Printed on permanent acid-free text paper.

This book is dedicated to
my wife, Anja, without whom it would not exist
and to all those who shared their work and teaching experiences with me,
both inside of and outside of medical education

Contents

Note of Thanks ... 9

List of Figures ... 11

List of Abbreviations ... 13

1. Introduction: Professionalisation of Medical Education ... 15

2. Profession and Institutions ... 21
2.1 Professions ... 21
2.2 Institutions ... 31
2.3 Summary ... 37

3. The German Medical Educational System and Simulation-Based Training ... 39
3.1 German Medical Educational System (MES) ... 40
3.2 Design of German Medical Studies ... 47
3.3 Simulation-Based Training (SBT) ... 52
3.4 Summary ... 60

4. Study Design: A Mixed-Data Approach ... 63
4.1 Methodology ... 64

4.2 Procedure of the Present Study 70
4.3 Summary 84

5. An Analysis of SBT's Three Arenas 85
5.1 Legal Arena: Overview of Formal Composition............................. 86
5.2 Public Arena: Issue Making of SBT.. 98
5.3 Workplace Arena: Case Studies of SBT in North-Rhine Westphalia.......... 121
5.4 Open questions towards triangulation 146

6. Triangulation of jurisdictional arenas..................................149
6.1 Incremental Change: Legal Versus public 149
6.2 Incremental Change: Legal Versus workplace 156
6.3 Incremental Change: Public Versus workplace 166
6.4 Summary 178

7. Conclusion and Outlook 181

8. Appendix - Available Digitally 197

References 199

Note of Thanks

A project of this size would not have been possible to realise without a supportive working environment. I would, therefore, like to thank my first and second supervisors, Thomas Heinze and Volker Remmert, for their constant support, helpful tips and critical feedback proffered throughout the project. I would also like to thank Laura Behrmann for our discussions about qualitative and educational research. My thanks also go to Anna Harris for giving me the opportunity to come to Maastricht and for her thoughtful and insightful feedback, offered from a medical educational standpoint. I would like to thank Maximiliane Wilkesmann for taking an interest in my project and for her encouragement. Finally, I would also like to thank all of the aforementioned individuals for taking time out of their busy schedules to serve as members of my PhD committee.

Furthermore, I would like to thank the Wuppertal GRK2696 community and its PIs for their company and valuable feedback provided throughout this journey. Special thanks go to my fellow PhD students, Jan Nicolay, Jeremias Düring, Sarwar Ahmed and Fabienne Grimm, and to the student assistants Verena Senger, Minori Nohara and Philine Higy. I would also like to thank the FASoS community at Maastricht University for their hospitality and open-mindedness. Finally, I would like to thank everyone who provided comments at conferences, workshops or other get-togethers that helped me to both think through and to work on this dissertation down through the years.

A project of this size would also have been impossible without the support of my personal network. All of my family members and friends have now heard about Simulation-Based Training, due to the many questions asked and because of their open ears, as well as my ramblings, whether they wanted to or not. Conversely, as a sociologist interested in professions, I have never liked the term 'work-life balance', as it implies that work is somehow separate from our lives. However, there have been many occasions when it has been wonderful to forget about my PhD work for a while. For this, and for your constant support, I am forever grateful. Luckily, there are too many of you to list, but you know who you are!

List of Figures

Figure 1: Professionalisation – Own representation after Abbott (1988) 43
Figure 2: Own representation of professionalisation of medical education.... 44
Figure 3: Overview of interviews 76
Figure 4: Three Approaches to Content Analysis (Hsieh & Shannon, 2005, p. 1286) 79
Figure 5: Congratulation to the founding of the JME (1984-H1 (9))........... 100
Figure 6: Activities of the GMA Committee Practical Skills (2011-V28-I4 (9), p. 9) 112
Figure 7: Rise of the term Skills Labs 119
Figure 8: Overview of the interviewees' educations..........................123
Figure 9: Simulation centres in Austria, Germany and Switzerland (Fandler, 2012, p.23) 168

List of Abbreviations

BÄK	Bundesärztkammer
BMG	Bundesministerium für Gesundheit (engl.: Federal Ministry of Health)
DRG	Diagnosis-related-groups
EPA	Entrustable Professional Activities
GMA	Gesellschaft für Medizinische Ausbildung (engl.: Association for Medical Education)
HI	Historical Institutionalism
IMPP	Institut für Medizinische und Pharmazeutische Prüfungsfragen (engl.: Institute for Medical and Pharmaceutical Exams)
JME	see below: ZMA
KBV	Kassenärztliche Bundesvereinigung (engl. Associations of Statutory Health Insurance Physicians)
MES	Medical Educational System
MFT	Medizinischer Fakultätentag (engl.: Standing Conference of Medical Faculties)
NKLM	Nationaler kompetenzbasierter Lernzielkatalog Medizin (engl.: National Competence-based Learning Catalogue for Medicine)
NRW	North Rhine-Westphalia, one out of sixteen Länder of Germany
OSCE	Objective Structured Clinical Examinations
SBT	Simulation-Based Training(s)
SP	Simulated Patients/Standardised Patients
ZMA/ JME	Zeitschrift für medizinische Ausbildung/Journal for Medical Education

1. Introduction: Professionalisation of Medical Education

Medical education is an integral part of the medical profession and involves training future medical professionals. The *professionalisation of medical education* is exemplified by the *rise of Simulation-Based Training*, which this study showcases as a distinct transformation. A historic-sociological analysis of this transformation *reveals incremental change* in a seemingly stable field, the advocacy of *new cognitive and material innovations* and the establishment of institutional structures that reinforce the *adoption of a new educational methodology*. Simulation-Based Training (SBT) is a new approach to practical training in medical education that is based on *competence-oriented pedagogy* and that introduces a distinct spatial and material organisation in *institutionalised training spaces called Skills Labs*.

The present study focuses on Germany, and on the subpart of North Rhine-Westphalia more specifically, in order to showcase both SBT's rise since the 1980s and its subsequent institutionalisation at German medical state-faculties since the 2000s. These developments are understood in terms of a professionalisation of medical education in Germany through both *cognitive and social claims of professional jurisdiction*. The present study's objective is to ascertain the extent to which relevant agenda-setting, by stakeholders in the respective professional arenas, has contributed to SBT's acceptance and institutionalisation, as represented by the founding of Skills Labs.

Two theories were used to examine the aforementioned professionalisation process. The first theory employed is the sociological theory of professions, as set out by *Andrew Abbott* (Abbott, 1988; 1991). According to Abbott, the importance of professions lies in their ability to apply abstract, academic knowledge to concrete problems. These entities engage in a battle for a monopoly in the context of *professional competition*, a strategy aimed at consolidating both their position and safeguarding their interests in professional arenas in the long term. Professionalisation is, therefore, the process by which a professional (sub)group claims professional jurisdiction (Abbott, 1988, p. 97). The second theory employed is Historical Institutionalism (HI), a term coined by *Kathleen Thelen* (Streeck & Thelen, 2005; Thelen, 1999). Thelen's theory, originating in political science, explores the transformation of institutions *as incremental change* that is based upon both external and internal decision-making processes and the competition of stakeholders over time. Different formats of institutional change exist, such as Layering, Displacement, and others (Streeck & Thelen, 2005, pp. 19–31). Consequently, a more extensive timeframe is required for a historical institutional analysis in order to avoid overlooking both continuities and discontinuities that might otherwise get taken for granted.

The analysis is based on three data types: First, there are the *changes in legislation* since the 1990s that took place via the national examination regulations and the curricula of all thirty-seven German state universities. The second data type includes the German-speaking *public expert discourse* via publications in the GMS Journal for Medical Education, dating back to its foundation in 1984. The third data type concerns case studies drawn from the contemporary 2020s *working modalities*, with regard to personnel structure and the commodities of Skills Labs, examined through expert interviews with medical educators and participatory observation of SBT in North Rhine-Westphalia's eight state medical faculties. All three data types have been analysed using content analysis and have been triangulated with each other in order to provide a thorough historical and contemporary picture (Hsieh & Shannon, 2005; Schneijderberg et al., 2022).

The institutional analysis of the transformation gives rise to the primary question: How can the rise of Skills Labs at German universities be characterised? Various answers to this question can be hypothesised.

One hypothesis suggests that the implementation of Skills Labs represents Layering of existing medical training techniques, thereby indicating an enhancement of these methodologies. Alternatively, it could be viewed as a Displacement of prior medical training methodologies, suggesting a substitution of such techniques in this case. Finally, it could be argued that the phenomenon under discussion is best explained through a combination of both Layering *and* Displacement. The final hypothesis is substantiated by this study's findings.

The transformation observed is neither complete nor is it universal across all levels. Rather, it is imperative to differentiate both the institutional restrictions of formal legislation and structure and the informal acceptance of newly implemented concepts at each individual location over time through teaching and testing. However, three distinct stages can be outlined to contextualise the general rise of SBT and Skills Labs. First, there was the pioneering phase, which included public advocacy for SBT, self-built materials and provisional simulation rooms as predecessors of the Skills Lab; this took place between the 1980s and up to 1998. Secondly, there was the establishment phase, spanning from 1999 to 2011, which saw the formal establishment of Skills Labs, the very first of which was founded in Berlin in 1999. The debate on competence-oriented education and the training of practical skills also took centre stage during this period. This resulted in the quasi-legislation of the NKLM (National Competence-based Learning Catalogue for Medicine) in 2012; a consolidation phase of Skills Labs can be observed happening from 2012 into the 2020s. At this stage, the question is no longer directed to whether medical education in Germany should consider SBT and the establishment of a Skills Lab at faculties, but instead to how their standardisation and quality criteria might be ensured. The case study of the eight state-run medical faculties in North Rhine-Westphalia illustrates the diversity of contemporary SBT implementation; this diversity in approach is due to the faculties' autonomy and variegated emphasis on medical education. The views of medical educators, who are involved in SBT for teaching and testing at present, shed light on their efforts to facilitate the transfer of SBT knowledge to clinical practice. These efforts are evident in the formulation of learning objectives, consideration of SBT fidelity and in terms of the management of Skills Lab spaces and equipment. These efforts serve to reinforce the long-standing tradition of both tinkering and in the gradual professionalisation that can be attributed to SBT more generally: "This pattern of power assumes, correctly, that jurisdictional invasion generally begins in the workplace, then moves to the public mind, and then into the law." (Abbott, 1988, p. 139).

Furthermore, this development in medical education cannot be understood in isolation. Like many other healthcare systems in the Western world, the German system is under pressure due to political and social changes (Busse, Blümel, Knieps, & Bärnighausen, 2017, p. 883) and this most notably due to economic measures and austerity measures introduced in the mid-20th century (Illing, 2022; Janda, 2022, pp. 35–36). These changes affect both the working and clinical training conditions of future medical professionals, due to reduced opportunities to interact with patients as a result of shorter hospital stays (Braun, 2014, p. 95). However, the pressure on clinical staff is even greater, resulting in different coping strategies being adopted (Molzberger, 2020), most notably manifested in a reduction in the incentives to teach novices (Wilkesmann, 2009, p. 267).

The reaction from the medical education sector has been, and continues to be, an advocacy for training opportunities for practical skills for which SBT is a new methodology. Both SBT's rise and the introduction of Skills Labs must, therefore, be contextualised in terms of the pressure being exerted upon on contemporary healthcare systems, as well as within the framework of new theories relating to competence-oriented education (Miani et al., 2015, p. 101). With regards to practical training in particular, traditional educational concepts encourage students to become self-sufficient as quickly as possible. This notion is often encapsulated by the phrase *'see one, do one, teach one'*, a view which has been criticised as potentially endangering patients (Singh et al., 2013, p. 13). Instead, the concepts of reform pedagogy and competence-oriented education advocate shifting the focus away from high-pressure performance assessments of students towards supporting their learning experience and identifying performance gaps. This approach seeks to produce more competent students and, consequently, future professionals, while also enhancing patient safety (Fabry, 2022). Simulation-Based Training (SBT), which includes teaching and testing with dummies, models, actors and other didactic commodities, offers one such safe space in which to develop practical skills, according to advocates (Owen, 2016). SBT's not-real learning environment has also attracted criticism for potentially producing skills that cannot be transferred to real clinical environments, as well as for medical education's increasing biomedicalization (Johnson, 2007; Nott & Harris, 2023; Prentice, 2013). In any case, SBT's increasing importance in medical education cannot be explained by one factor alone, but should instead be considered alongside the parallel development of various incremental transformations that unfolded throughout various professional arenas. SBT's design requires dedicated spaces; hence, Skills Labs were developed as facilities for practical training.

The extended research questions, therefore, include: how and why did these educational developments take place? To what extent have Skills Labs been implemented at faculties to date? And what are the medical educators' stances on this shift in practical medical education? This book's aim is to provide a multi-layered analysis of the incremental changes made to medical education in Germany since the 1980s. This will make the professionalisation of medical education through SBT visible and will enable the analysis of different professional arenas in which stakeholders assert their claims as competing, professional subgroups (Abbott, 1988, p. 97).

This study will yield insight into various research areas. On the one hand, the study is characterised by its strong references to the *sociology of professions*; on the other hand, it also offers a *historical institutional analysis*, which is novel, by dint of this combination. Furthermore, the research subject and its technological development also touch upon areas of *science and technology studies* and *educational research*. SBT's rise exemplifies the evolution of a well-established profession and professional subgroup, such as medicine and medical education, thereby indicating that fundamental (albeit just incremental change at times) is transforming the entire profession profoundly. This study will be of particular interest to those who are researching professional development and transformation, those researching medical materiality, educational or institutional structures as well as to stakeholders in the fields of either medical or higher education.

2. Profession and Institutions

Two frameworks of theories, originating from different scientific backgrounds, inform this study. The first theory is the sociological theory of professions; this mainly draws upon Andrew Abbott's work. The second theory is the political science theory of institutions and mainly follows Kathleen Thelen's work. Both theories present beliefs about competition between social groups as well as (incremental) processual change. Both thinkers' works can and have been implored in the past to better understand empirical objects, such as knowledge production and the changing of institutions (Hallonsten & Heinze, 2013), professionalisation and the academic system (Jappe et al., 2018), technological transformation and professionalisation (Petersohn & Heinze, 2018) and the medical profession (Oh, 2014). However, combining both theories is a novel approach. To that end, both theories will be introduced and will be applied to the subject of medical education and SBT in turn.

2.1 Professions

> "The tasks of professions are human problems amenable to expert service" (Abbott, 1988, p. 35).

This study's main reference for the framework on professions and professionalisation is Andrew Abbott and his book "System of Profession" from 1988.[1] Abbott, a US-American author, refers to a number of international, historical and – back then – contemporary examples in order to compose a unified sociological theory of professions. As outlined in the quotation provided above, professions solve individuals' problems through their special knowledge and, therefore, form a group of experts who are responsible for such solutions. This comprises their *professional jurisdiction*. For this study, this concerns the subgroup of medical education, from the classical profession of medicine, and its novel methodology of Simulation-Based Training. Simulation-Based Training, in turn, is defined as the professional practice of medical educators to utilise dedicated spaces and commodities aligned with their educational goals in order to teach professional novices (i.e., students of medical studies) clinical skills and competencies.

According to Abbott, professional practice consists of three steps: *diagnosis, inference* and *treatment.* Diagnosis involves defining the features that are important for the diagnosis of the professional problem. Inference involves recognising the different, possible solutions to the professional problem and deciding upon the best one. For Abbott, this is what true professionalism entails. Lastly, treatment involves the professional implementation of the solution (Abbott, 1988, pp. 40–52). In another passage, Abbott draws parallels between this three-step system and the complexity of chess. He asserts that while the opening and ending of professional practice are often formulaic, the middle part (the inference) represents strong problem-solution skills (Abbott, 1988, p. 48). The transfer of this concept implies that the necessity for practical medical training and the execution of SBT are comparatively less

---

1 Naturally there are numerous other authors who could have been considered for their contributions to the sociological theory of professions and, to a more specific extent, the medical profession itself. However, due to limitations of space, these authors cannot be introduced in this context, only deserving a brief mention (i.e., Becker et al. (2007); Dick (2016); Foucault (1973); Freidson (1995; 2001); Goffman (2016); Hughes (2015 – Orig 1958); Larson (1977); Nettleton (2021); Parsons (1939); Pfadenhauer (2003); Stehr and Grundmann (2010) and Timmermans and Berg (2003a).

complex than the definition and implementation of SBT as a training methodology (i.e., by specialised knowledge).

Professional practice takes knowledge, gained by the professional through their *academic education*, and is characterised as *specialised, abstract knowledge* (i.e., the knowledge of didactical theories and methodologies in SBT (Abbott, 1988, pp. 52–57)). However, professions establish both cognitive and social claims because there are different possible solutions to a given problem and there are competitors for optimal solutions, as indicated previously. The *cognitive claims of a profession* regard the abstract, specialised knowledge and are, therefore, connected to the academic sector that both produces and refines the discipline's knowledge. In this study's context, a concept's application to education, such as competence, is examined in relation to the specific professional field of medicine through the use of SBT. A profession's *social claims* are defined according to the cultural context of its practice or organisation, as evidenced by the adoption of competence as the leading theoretical concept by both individual universities and study programmes (Abbott, 1988, p. 58).

Accordingly, any profession is situated between the abstract knowledge, upon which it is based, and the social circumstances in which this knowledge is applied to concrete problems. The individual professional must, therefore, continuously balance *optimum abstraction*, keeping up to date with the latest academic findings while selecting those that are most useful for their professional practice (Abbott, 1988, pp. 105–106). The very same balance is also necessary for the education of any profession, due to the very abstract nature of academic education in comparison to the professional practice of working conditions the future professionals have to emerge in, because "In reality, this perfected abstract knowledge system exists only in the world of professional textbooks" (Abbott, 1988, p. 56) to which both teachers and students refer. This regards the educators of Skills Labs, as leading examples of SBT's implementation, in specific.

Professional Arenas

According to Abbott, the claims to jurisdiction by a profession can be established in three different arenas: First, there is *the legal arena*, which concerns the laws, regulations and official side of institutions that the professions want to shape according to their claim of responsibility. Second, there is *the public arena*, which includes the agenda-setting and discourse on professional issues (i.e., in (specialist) media, public statements, et cetera). Third, there is the *workplace arena*, which relates to the professional practice, as described previously, and the social networks of professionals to one another, if applicable. The relationship between the different arenas can be complimentary, insofar as the legal system ensures formal control over the workplace, while regulations can be circumvented in the workplace arena. The legal system provides discussion content for the public arena, while the public arena can pressure the legal system towards adopting different legislation. Finally, the public arena and workplace arena can re-inform each other about working conditions, issues and needs relating to ongoing professionalisation (Abbott, 1988, pp. 59–68). This analysis uses three types of data in order to represent medical education's three professional arenas. These include leading institutions and regulations, public discourse in scientific articles, and organisational circumstances and commodities surrounding SBT and Skills Labs (see Chapter 3).

To return to the argument regarding the basic definition of a profession, outlined previously, while this social and cultural authority is a deliberate aim for any profession, the superiority aspiration or the *"fight for turf"* (Abbott, 1988, p. 102), is rooted in specialised knowledge claims. In short, a given profession might argue that because they are in the possession of 'better', 'higher', 'more applicable' knowledge, then they can claim jurisdiction. Applying this argument to this study's subject, the authors of medical education literature explicitly ground their concept of competence, and the transfer of professionalism, in the specific, abstract knowledge of extended pedagogical theory (see Chapter 3.2 for a more detailed discussion of competence definitions). However, medical education is subordinate to the broader medical profession, which requires a settlement between professional groups (Abbott, 1988, pp. 68–69): "It is an inherently uneasy settlement, partly because it is undercut by workplace assimilation and partly because subordinates become absolutely necessary to successful practice by superordinates" (Abbott, 1988, pp. 72–73). Lamentably, settlements of this type within one profession not described by Abbott at length; instead, he often focuses on settlements between different professions and that can entail even higher competition (e.g., with nursing being a subordinate profession to medicine) while continuously fighting for the enhancement of their position (Abbott, 1988, p. 71).

However, this thesis is not a comprehensive study of the medical profession; rather, it is an examination of medical education that focuses on SBT. Medicine's general purpose is to heal; for medical education this involves the transfer of medical knowledge for future professional practice. It is necessary, therefore, to consider the external environment and the internal structure of medical education as a professional group, and how it professionalises, in order to analyse these claims of jurisdiction in the three arenas mentioned previously (see Chapter 3.1).

Professionalisation

Changes towards a system of professions, and in its division of labour, can have either external or internal origins. External factors can include changes in technology or to the organisations surrounding the profession, while internal factors can include organisational development and professionalisation. These changes can result in a new constitution of professions and in an alteration to the strength of their jurisdictional claims (Abbott, 1988, p. 111). A subordinated professional group can "move to higher branches" through professionalisation, which is considered an internal source of professional system change (Abbott, 1988, p. 97). This professionalisation is attributed to medical education through SBT's implementation as methodology and to the overarching didactical concept of competence, but a thorough, processual analysis is required in order to fully understand the extent to which these factors have contributed to the developments observed (see Chapter 6).

The notion of change, commonly beginning with small incidents, is very much in line with the notion of incremental historical change that historical institutionalism argues for (see Chapter 2.2.). The composition of a profession's *internal structure* and its social stratification is defined by the composition of *groups* within that profession (i.e., strong or weak social connections between professionals leading to a coherent development of claims) and relating to the *control* that the profession possesses (i.e., the rules and regulations the profession actively shaped) and, lastly, the worksite that considers the *work organisation* (i.e., union and communication at the smallest level (Abbott, 1988, pp. 79–85)). This study's internal structure will outline how medical educators stand their ground against clinical teachers and other stakeholders, specifically regarding SBT's organisation and management and its commodities within Skills Labs as 'their' institutions.

Furthermore, as in any social group, professions also indicate *stratification* and, once again, Abbott hints at the root of professions within specialised knowledge. A profession's subgroups that perform "in the most purely professional environments" (Abbott, 1988, p. 118) are attributed a higher status by their peers and by the public due to their closeness to the given 'pure' academic knowledge. Changes in status can function as an indication of change for the entire profession, emerging from outside or from inside, especially because "jurisdictional invasion generally begins in the workplace, then moves to the public mind, and then into the law." (Abbott, 1988, p. 139), as is visible in medical education.

To conclude, Abbott attributes professions a social monopoly towards specific social problems because of their specialised knowledge and *professionalisation as the process* of external occupations or by internal subgroups claiming jurisdiction; this can be analysed within three arenas (legal, public, workplace), as well as through changes made to either internal structure or on the basis stratification, as the profession's reaction to challenges (Abbott, 1988, p. 193). The medical profession is being transformed by the professionalisation of its subgroup in the field of medical education; this is claiming jurisdiction in the legal, public and workplace arenas in order to introduce and consolidate SBT as a training methodology.

While an extended examination of the debate on the literature regarding professions, which would augment Abbott's idea, is impossible due to space limitations, the idea of professionalisation has been incorporated into most theories written more recently. A general definition, provided by Evetts (2013), captures its essence:

> "The concept of professionalisation is regarded as the process to achieve the status of profession and has been interpreted as the process to pursue, develop and maintain the closure of the occupational group in order to maintain practitioners' own occupational self-interests in terms of their salary, status and power as well as the monopoly protection of the occupational jurisdiction" (Evetts, 2013, p. 782).

Professionalisation also has an endpoint, as a process, which culminates in *professionalism*. Professionalism, and thus the embodiment of expertise about specific societal problems, is represented in three ways. First, it is represented by the individual's professional practice and by their inherent specialised knowledge. Second, it is represented by the division of expert labour in the form of an internal structure. Finally, it is represented by a profession's *commodities* (Abbott, 1991, pp. 21–28). The process of professionalisation in medical education is still underway and, thus, no professionalism has been reached yet. Nevertheless, the indicators of higher professional standards can nevertheless be considered.

Commodities include any material embodiment of specialised knowledge, insofar as this knowledge was necessary to producing them and they are integral part of the professional practice. In the case of the medical profession, this includes the index of classification of diseases, thereby reflecting its historical evolution and the ongoing debate surrounding the classification of diseases. Moreover, a multitude of instruments and technologies that are central to the practice of medicine can, thus, be considered commodities. Some, such as the stethoscope or an X-ray, are more directly pertinent to diagnosis, whereas computer programmes for the organisation of patient files are less directly related to the diagnostic process, but the computer itself is not a commodity specific to the medical profession. Nevertheless, all of the examples mentioned previously can still be understood as commodities according to Abbott (Abbott, 1988, p. 324, 1991, p. 21). Commodities can be put into three mayor categories for medical education and for SBT in specific: simulators, simulated patients and materials on the verge of SBT (see also Chapter 3.3). Finally, it is essential to highlight certain, distinctive features of the medical profession generally in order to better situate medical education within it prior to exploring the theory of institutions as the second major theoretical framework that informs this study.

Medical Profession: For the People and of the People

The abstract knowledge, described previously as being central to professionalism, is not simply handed down. Abbott's concept of an 'academic system' is one in which abstract knowledge is both developed and obtained. However, it is important to recognise that each profession and discipline has its own historical transformation. Medicine's transformation is marked by the development of distinct societies with unique political and intellectual systems, accompanied by the emergence of knowledge transfer and storage mechanisms, such as monasteries and universities (Land, 2018, pp. 16–17). The roles played by doctors and patients, the concepts of health and sickness, and the scientific – or nowadays considered pseudo-scientific – systematisations of bodily functions have all been influenced by these developments (Duffin, 2021; Jutel & Nettleton, 2011; Kögel, 2021, pp. 15–19). In the present study's context, the claim of expert knowledge and its organisation must be theorised and accepted in its simplest form, as described by Abbott.

However, there are divergent, critical approaches to the theoretical representation of these developments. David Armstrong, for instance, posits that the analysis and objectification of bodies and patients' histories is rooted in the discipline of modern societies, thereby building on Foucault's theories (Armstrong, 1983, p. xi). Others retrace how the professionalism of the modern health care system became reframed towards a clinical rationale (Gradmann & Simon, 2010). These developments can be best described as biomedicalisation (Nettleton, 2021, pp. 2–5; Prentice, 2013, p. 17; P. H. Roth, 2021). This study aims to reflect the rise of SBT's educational methodology through formal and informational jurisdictional shifts, given that there are authors who draw the correct connection that medical education also teaches biomedicalisation and should reflect upon its knowledge organisation and perception (Johnson, 2007; Nott & Harris, 2023). The study functions as an inventory and as a warranty, exploring the mechanisms that enable both dominance and subordination.

Furthermore, and like with medical knowledge, the medical professional practice is no exception in the linking of different levels of both formal and informal change, since every medical professional is bound by the formation of a given society, be it through individual morals and social conventions or due to the health systems and economies of which they are a part (Armstrong, 1983; Karsch, 2015; Temkin, 1949). Medical sociology, and its connection to science and technology studies, has espoused a number of coherent theories and has examined particular study objects throughout its rich development during the 20th and 21st centuries (Casper & Morrison, 2010; Nettleton, 2021; Stollberg, 2001; Timmermans & Berg, 2003b). This tension, as represented in the literature on the sociology of medicine, must be understood as existing across the entire profession. Individual professionals must make personal decisions, regarding the manner in which they treat their patients and the aspects of the system that they find problematic, in the context of a healthcare system characterised by economic efficiency. Analyses of these divergent expectations and strategies illuminate the professional practices in question (Atzeni, 2016; Becker et al., 2007; Molzberger, 2020) and can also be found within this study (see Chapter 5.3 and 6).

One example of a contemporary analysis of medical and professional organisation, which also deploys Abbott's theory of professions, is the study by Hyeyoung Oh from 2014 (Oh, 2014). She describes the claims of jurisdiction and intra-professional fights surrounding consultations within US-hospitals on the basis of ethnographic observation and interviews. Oh deduces how the medical specialties are attributed different status', and how this stratification influences patient case management, by attending patients rounds and internal meetings, in addition to statements made by the professionals themselves. Oh concludes by claiming that formal knowledge does not commonly serve as the basis for decisions, especially because not all information can be gathered in short time spans, but rather that the formal hierarchical status of career and informal hierarchical status of importance of specialties inform the decision-making process (Oh, 2014, pp. 580–586). Three major strategies can be seen to be at work here, according to Oh's study. Either the case management is clearly, based on the facts provided, then the academic knowledge is a contender for a decision, or the informal collegiality (or rivalry) towards another specialty is pivotal, in which case the social bonds are the most influential, or a higher hierarchy is taken into consideration, in which case power is the social mechanism that is employed (Oh, 2014, pp. 589–592). On the one hand, Oh's study demonstrates that Abbott's theory can be readily reproduced and expanded into examinations of contemporary conditions in the health care system, as evidenced by the US-American case study. On the other hand, Oh also demonstrates that the competition and "imperialism" of the professional jurisdiction inherent to Abbott's theory may not be the only strategy or mechanism in operation. The dynamics of cooperation and rivalry, which Abbott subsumes under the rubric of internal structure, are also at play and should be given careful consideration when examining the organisation of professional groups within the health care system. Thus, the different status groups within medical education, which either hinder or support the rise of SBT, will also be closely analysed within this study (see Chapter 5.3).

To conclude this short overview of medical sociological approaches, Nettleton, inspired by Seale, underlines differences between studies *in* medicine and *of* medicine and health. Studies in medicine and health consider the practices of a given health care system on individual, social and societal levels, thereby representing a micro-, meso- or macro- perspective. However, studies of medicine analyse health care systems and its subparts with regard to their mechanisms (i.e., how disease is socially constructed, how dominance is established and what influence different political systems play (Nettleton, 2021, p. 9)). This is a study *of medicine* on a *social level* and it considers the power dynamics at work within the medical profession, in this case focusing on the subgroup of medical education in order to better understand its increasing domination and diverse jurisdictional claims. While this study cannot (and will not) describe neither the historical variance of medical knowledge nor medical education overall in detail, it will instead specialise on one medical educational procedure and its tools – SBT, its institutions and commodities.

The preceding sections have served to mark out the field of medical sociological analysis generally. The one detailed study by Oh is particularly interesting, due to the successful use of Abbott's theoretical framework, something that is also considered in this study. As a concluding remark, later work by Abbott (Abbott, 2001; 2020) underlined his process-related thinking to an even greater extent than the numerous examples of historical changes that Abbott describes in "System of Profession" (Abbott, 1988). Therefore, his analysis asks for a sociological theory of professions and knowledge systems based in historical analysis and it offers (correspondingly suitable) theoretical concepts (Hoebel et al., 2020). A second branch of social theories, namely that of institutions, are used to further outline and supplement the social forces that are deemed central for the present research and this requires a focus on institutions and a focus on different levels of analysis (i.e., from meso- to micro- level (see Chapter 3)).

2.2 Institutions

Connection of Professions and Institutions

Abbott states, in the beginning of System of Professions, that while he was originally only going to consider one profession, he found that his broader theory of professions must consider institutionalisation, making professionalism one form of institutionalisation (Abbott, 1988, p. 9). While Abbott addresses the institutional structures of professions (i.e., their educational systems and licensing procedures), his book does not specifically address the question of what exactly constitutes institutions. His later work defines institutionalization (sic) more precisely:

"At the simplest level, institutionalization of expertise means the emergence of a set of rules for handling it, [...] and arrangements of those norms and roles into larger structures – organizations for delivering expert services, hierarchies of types of expertise, routines for reproducing expertise, and so on" (Abbott, 1991, p. 20).

This definition outlines how informal formations, consisting of expertise towards a specific subject, become formal in form of rules and organisations, "handling" the given profession and "reproducing expertise". The Skills Labs (defined more exactly elsewhere, see Chapter 3.1) and their introduction to medical faculties represent one such case of formalising informal interest in this new methodology of SBT into institutions of their own, being shaped towards their needs and both providing and reproducing the expert service of medical education in terms of practical skills training.

In the vein of Abbott's interest in the reproduction of expertise, institutionalism is particularly prominent in both organisational sociology and within political science. Authors within this sphere have remarked that (expert) occupations as topics oftentimes, if not practically always, ask for the consideration of institutions with respect to a given occupation's two-sided nature of "institutional systems constraining and empowering individual participants" (Scott, 1995, p. 9). Historical Institutionalism (HI), this study's second theoretical pillar of medical education, allows for a two-sided analysis and encompasses a specific language for institutional transformation over time.[2] In SBT's

2 Other formats of institutional theories cannot be considered in detail, due to space limitations; however, three basic schools can be differentiated: Rational Choice Institutionalism, Sociological Neo-Institutionalism and Historical Intuitionalism. Rational Choice Institutionalism is based on the idea of utility maximisation. This means that both actors and institutions continually base their actions on enhancing their own interests in a rational manner, according to Greif and Laitin (2004, p. 635) and Weingast (2002). Sociological Neo-Institutionalism considers cultural circumstances to be a necessity in order to have actors and institutions evolve (Scott, 1995, p. 33), while the "(...) institutional arrangements provide the cultural milieu that determines what kinds of actors can exist, what they can do, and what meanings their actions have" (Scott et

case, the Skills Labs bundle the institutional forces of personnel, space and commodities, as well as content-related and financial support by their mother institution, the universities.

HI is also occasionally defined as political institutionalism, given that it combines notions of the 'homo oeconomicus' and the 'homo sociologicus' by including the rational decision-making processes, while accepting that actors are neither omniscient nor are they resistant against exercise of power (Wilkesmann, 2009, pp. 126–127). The central representative of this strand, and the main theorist for this study, is Kathleen Thelen. Her work has previously been applied to describe changes in scientific institutions (Aagaard, 2017; Heinze & Münch, 2012) and can also be expanded to include other types of organisations. It considers not only political entities, but also any regime-building structure within public society (Streeck & Thelen, 2005, p. 12):

"Very generally, institutions may be defined as building-blocks of social order: they represent socially sanctioned, that is, collectively enforced expectations with respect to the behavior of specific categories of actors or to the performance of certain activities. Typically they involve mutually related rights and obligations for actors, distinguishing between appropriate and inappropriate, 'right' and 'wrong', 'possible' and 'impossible' actions and thereby organizing behavior into predictable and reliable patterns" (Streeck & Thelen, 2005, p. 9, Italics in original).

The collective enforcement of which institutions are capable provide metaphorical guard rails for actors within an institution. They leave room for strategic behaviour, such as utility maximising behaviour, while also signalling that institutions have social forces that exceed

al., 2000, p. 363). In related theories, disputes are conceptualised depending on power and institutional asymmetries; this is exemplified by the symbolism of ideas and discourses according to Blyth (2002) and Schmidt (2008). Finally, historical institutionalists focus on the formal aspects of institutional composition with a special interest in temporality. Nonetheless, some degree of convergence can be formulated for all three strands: "Common denominators of institutional analyses are an interest in persistent, binding, and essential social forces and inquiries into how such forces remain stable or evolve over time" (Lohse, 2024, p. 88).

those of the individual (Streeck & Thelen, 2005, p. 11). The following section addresses the nature of these institutions' forces.

Institutional Forces

A crucial issue in institutional theory concerns the relationship between an institution's formal structure and the possibility to overthrow it, thereby causing significant transformation. This is a question concerning an institution's reproduction and the possibilities for long-term reorganisation. On the one hand, institutions hold the possibility to evolve, especially through the introduction of new educational backgrounds for the agents within institutions. On the other hand, though, many institutions and their internal rules can also be described as sticky. This stickiness is attributed to the constant reproduction of the given rules and this helps to explain their relative stability (Thelen, 1999, pp. 386–387). Consequently, institutions are frequently regarded as 'locked-in' to their processes, a phenomenon known as path dependency (Pierson, 2000; 2015). However, path dependency has limited analytical value in the case of SBT because it does not demonstrate how or why stakeholders could or should introduce new concepts or oppose the designated 'winner' of an argument. It does not explain how or why educators might reorient themselves towards competence-oriented education and SBT's methodology. Overall, SBT demonstrates medical education's inertia and gradual transformation, whereas the implementation of Skills Labs at German medical faculties represents a significant transformation.

Consequently, HI does not dismiss the concept of path dependency, but instead questions its exclusive explanatory capacity (Thelen, 1999, p. 385). This notion entails that each development and event play a distinct role in shaping the options available for action, thereby limiting the likelihood of certain outcomes in arguments and, consequently, of consolidating specific 'paths'. As a result, the necessity arises for a historical approach that both outlines and contextualises the events of interest, while disregarding influences that are deemed to be of lesser importance. Applied to the individuals in this study, the professionals with their practices have to move within such paths (i.e., the SBT educators have to act within the given health care and educational system (Abbott, 1988; Zorn et al., 2006)) and sometimes even around the given paths if, for instance, there is a grave difference between the legislative jurisdiction/institutional enforcement and the "workplace realities" (Abbott, 1988, p. 84). Accordingly, HI does not emphasise structure over agency or vice versa, but instead understands both as social forces that both restrict and allow for the changing of institutions as "[...] actors cultivate change from within the context of existing opportunities and constraints" (Streeck & Thelen, 2005, p. 19).

It can be concluded that neither social scripts and norms, nor rational choice, nor the path-dependency of mapped progress are the sole determinants of various institutions' forcefulness within HI (Streeck & Thelen, 2005, p. 11). Instead, institutions are understood as *regimens* that consist of *rule makers, rule-takers* and *rules* and their *unintended consequences*. Rule makers include stakeholders who design rules for an institution or an organisation that serve their own interest. They are cognitively limited and cannot foresee all of their rules' possible implications. Consequently, the rules themselves must be sufficiently general to be applicable to different circumstances, but must be precise enough to recognise their applicability. Conversely, the individuals charged with the implementation of these rules, the rule-takers, may interpret them strategically, thereby altering the rule's intended purpose, thereby leading to unintended consequences (Streeck & Thelen, 2005, pp. 13–15). The SBT educators are rule-takers, with respect to the bigger medical educational system, occasionally bending or redirecting rules. In relation to the Skills Labs, they can act partly as rule-givers towards their professional practice and space management.

An unintended consequence can mean that the implementation of a new rule, working objects, teams, etc. within an institution does not necessarily function as expected by the *designers*/rule makers (Thelen, 1999, p. 382). The advantage of institutions as regimes can be found in the multitude of change origins, due to the gap between the imperfect rule formulation and its strategic implementation after the fact (Streeck & Thelen, 2005, p. 14). In conclusion, it is important to note that an institution's regime is not exhaustive and is, therefore, subject to perpetual change. This quality makes temporality a central concern of institutional theory.

Temporality

Some authors have delved into the intricacies of temporal reasoning concerning significant events, often termed *'critical junctures'* or *'disruptions'* when considering the composition and transformation of institutions. These events have the potential to abruptly alter the prevailing structure of an arena or institution (Collier & Munck, 2022; Pierson, 2015). Such transformation can also be termed *'punctuated equilibria'* (Orren & Skowronek, 2018), which assumes that most political and public processes are grounded in the equilibria of their respective stakeholders, following their interest and keeping each other at bay in the process; only when such an equilibrium is disrupted, by the significant changes mentioned previously, can fundamental transformation occur. Such critical junctures can also be observed within this study's timeframe, in terms of the period from 1980s to the 2020s in the field of medical education in Germany with the NKLM's introduction and the outbreak of Covid-19. Nevertheless, despite their internal and external nature, these forces proved inadequate to account for the entirety of the transformation observed.

Instead, the argument can be made that gradual internal and external change culminates in long-lasting transformation. This approach based in HI and is advocated for by Thelen et al. (Mahoney & Thelen, 2010; Streeck & Thelen, 2005; Thelen, 1999) and is favoured in this study to better understand different forms of institutional change over time. The authors discuss changes regarding their *abruptness* or *incremental* progress as well as their results as *continuous* or *discontinuous* for institutions (Streeck & Thelen, 2005, p. 8). We can conclude that (dis)continuities, which might seem small or ineffective, can be nonetheless part of meaningful gradual transformation. The simultaneous duality of formality and practice, as outlined above, is also stressed in the wording of institutions as a *'regime'*: "The biases inherent in existing conceptual frameworks are particularly limiting in a time, like ours, when incremental processes of change appear to cause gradual institutional transformations that add up to major historical discontinuities" (Streeck & Thelen, 2005, p. 8). Recent accounts also argue in favour of a combination of analysis of incremental change and critical junctures (Lohse, 2024, pp. 95–98; Stark, 2018), which is followed in this study.

Five different forms of gradual, institutional transformation are defined: namely, "Displacement, Layering, drift, conversion, and exhaustion" (Streeck & Thelen, 2005, pp. 19–31). Either Displacement or Layering effects are considered as the most fitting descriptors for the change observed within medical education. *Displacement* describes cases in which actors within a certain arena gain more and more influence due to weak leading stakeholders, thereby becoming the new dominant regime. *Layering* similarly entails the use of an already existing structure and involves the addition of new parts or purposes to change that institution's intended design, while refraining from overthrowing the entire structure from the outset. For SBT, this would entail one of two possibilities: Firstly, all medical education could be reoriented towards practical, competence-oriented, simulation-based education, displacing other regimes in the process. Secondly, SBT could be a leading concept within the realm of medical education, alongside others, and the given regime could be transformed through this Layering. Consequently, a timeline outlining the transformation of medical education, especially with regard to SBT, is necessary in order to comply with "[...] historical institutionalism (which) views causality as contextual, the contextuality deriving from complex configurations of factors that become apparent through historical-comparative analyses" (Lohse, 2024, pp. 88–89).

2.3 Summary

In summary, the two theoretical frameworks of professions and institutions employed in this study offer distinct perspectives on the organisation, jurisdiction and power at play in the reproduction of work and education. The concept of professions is useful for both analysing professional jurisdiction and practice as abstract knowledge that can be applied to concrete cases, as well as for understanding the professional arenas in which professions operate to protect their problem-solving monopoly, grounded in special knowledge (Abbott, 1988). Institutions, conversely, are subject to change when stakeholders within seize the opportunity and incentive, especially within HI along with outside circumstances, to alter or reinterpret the established rules in a manner that aligns with their interests (Streeck & Thelen, 2005). Abbott and the HI supplement each other in this respect.

Cognitive claims are activated, which are not yet as formalised as the HI considers to be relevant, whenever the pedagogical-theoretical justification as abstract knowledge changes. However, Abbott demonstrates that this comprises the basis of professional jurisdiction. Accordingly, it is important to consider the application of abstract knowledge in this context, through journal articles that use competence-oriented theories to justify SBT, for example. Conversely, initially informal social activities are likewise relevant, given that they can lead to formal regulations and because they represent social claims. The NKLM can be used as an example for SBT. Educators were already organising statements, committees within associations, and preliminary definition work on the NKLM even before the publication of the NKLM was used as quasi-legislation. While the HI would consider this incremental change to be important, only its manifestation in the NKLM, as a catalogue of definitions, provides a clear, formalised expression of competence orientation. Abbott, conversely, emphasises the social claims for professional jurisdiction as a change in a profession's culture, whereby the educators themselves take responsibility for the creation of new guidelines. This study's aim examines such supplementary analyses based on the two frameworks.

In short, the present analysis' objective is to outline the demands and advantages that medical educators perceive within the transformation of/compliance with SBT as a leading regime over time, provided that the transformation can be considered to be complete. The subsequent chapter 3.1 will provide a comprehensive overview of the key stakeholders and relevant arenas for the professionalisation of medical education in Germany towards SBT.

public and workplace arenas are mutually informative about current developments, insofar as the former represents the latest issues and statements made within the public discourse, thereby providing insights into standards and aspirations for the workplace more generally. In turn, the latter may relativise these aspirations, considering them to be exaggerated and impractical (Abbott, 1988, pp. 59–68). Applied to the professionalisation of medical education using the example of SBT, different actors can be identified for each arena as stakeholders, as shown in Figure 2.

In the public arena, in accordance with the German federalism (see below), we find the 'Bundesärztekammer' (BÄK) and its subdivisions, which represent professional interests and can issue public statements. The origins of the Ärztekammern lie in historical guilds, with each Land having its own Landesärztekammer, the representatives of which are chosen to join the Bundesärztekammer. Furthermore, all medical professionals who have undertaken the state examination automatically become members of their respective Landesärztekammer (Wernitz et al., 2022, p. 206).

In addition, the MFT (Medizinischer Fakultätentag, engl.: Standing Conference of Medical Faculties) and GMA (Gesellschaft für Medizinische Ausbildung, engl.: Association for Medical Education), among other associations, also need to be considered. The MFT is the largest non-governmental association that directly affects medical education, given that it is the standing conference of all medical faculties and is, therefore, a key stakeholder. The MFT functions as a union of medical faculties, putting forward statements on behalf of university medicine. The medical faculties view the professionalisation of medical education as one of their primary objectives (Medizinischer Fakultätentag (MFT) Deutschland, 2013, p. 25). The GMA is an association spanning Austria, Switzerland and Germany and that aims to improve the teaching of the medical profession (Gesellschaft für Medizinische Ausbildung, 2021, §2, (2)). It holds several thematic subcommittees, some of which are highly relevant to SBT. In addition to statements made by the Ärztekammern and other associations, contributions to professional discourse from sources such as journals and individual experts must also be considered. The GMA has also published its own journal since 1984; namely, the Journal of Medical Education (JME). This publication organ is used by the GMA for scientific studies, reports on medical education and to find individual statements from educational experts.

The Federal Ministry of Health (Bundesministerium für Gesundheit, BMG) is institutionally the highest authority in the healthcare system within the legal arena, according to the federal government. In Germany, administrative responsibilities are divided between the Bund (federal government) and the sixteen Länder (sixteen individual federal countries), in the manner of German federalism. While the federal level (Bund) represents the entire nation, each individual Land (plural: Länder) represents a subpart, has its own regional government and governs in a manner aligned with the Bund (Blümel et al., 2020, pp. 6–7).

The Federal Ministry of Health (BMG) issues the national examination order (Approbationsordnung), which grants licensure to practice as a physician in Germany (Simon, 2021, p. 173). The sixteen Länder must agree on this examination order because their educational institutions, the universities' medical departments, administer the content of the examination order in teaching and testing (Busse, Blümel, & Spranger, 2017, pp. 61–62). Debates between the Bund and the Länder often centre on issues of liability and finance (Wernitz et al., 2022, pp. 200–201). The Länder agreed to finance the Institute for Medical and Pharmaceutical Examination Questions (IMPP) to organise and guarantee the correct progression of the state examination in several medical professions, thereby making it a joint institution of all of the Länder responsible for preparing and evaluating written examinations in the undergraduate education of physicians, dentists and pharmacists (World Health Organization. Regional Office for Europe & European Observatory on Health Systems and Policies, 2000, p. 32).

Figure 1: Professionalisation – Own representation after Abbott (1988)

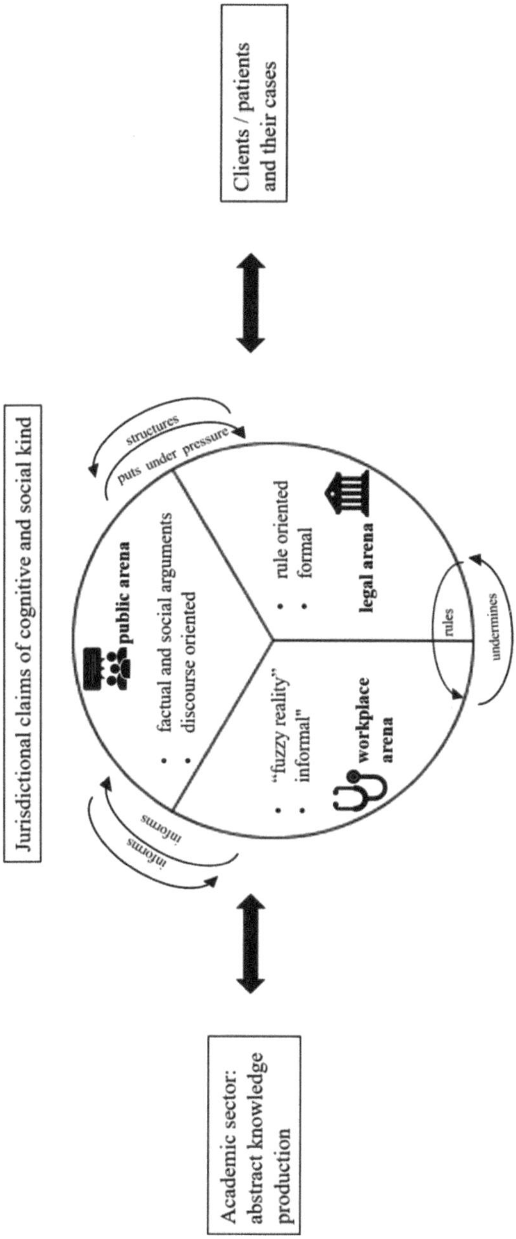

Figure 2: Own representation of professionalisation of medical education

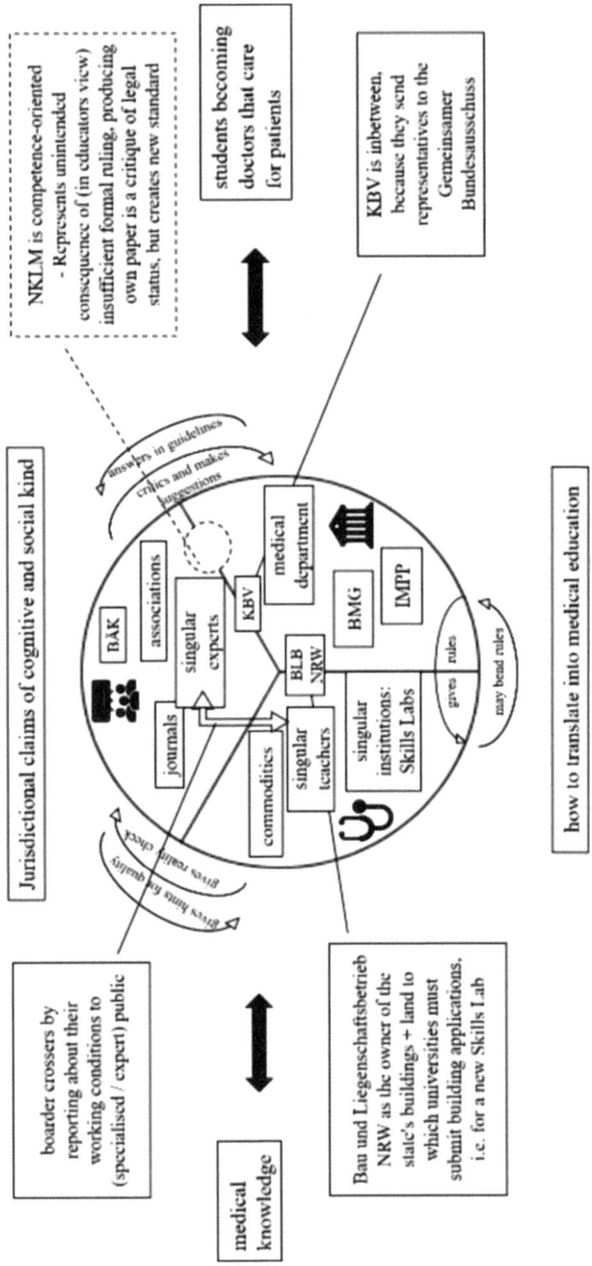

The universities' medical departments are divided into two structures: the medical faculties and the university clinics. The medical faculties are those academic entities that are responsible for both structuring and conducting medical studies. The university clinics, also known as teaching hospitals, offer internships for medical students and act as providers in the health care system. Personnel at the medical faculty and university clinic may be identical, insofar as an individual holding a chair at the medical faculty often also the head of a sub-division of the clinic. Most medical faculties in Germany, as well as all cases in North Rhine-Westphalia, follow this cooperative approach between the faculty and the university clinic (Medizinischer Fakultätentag (MFT) Deutschland, 2013, pp. 8–9).

The Association of Statutory Health Insurance Physicians (KBV) is another institution that operates across borders. Positioned between the public and legal arenas, it participates in public discourse by making statements and by representing the profession. It also has direct influence on legislation through the Joint Federal Committee (Gemeinsamer Bundesausschuss). The healthcare system is the primary focus of their concerns, although they do occasionally provide direct comments about medical education (Simon, 2021, p. 172).

The individual institutions that administer medical education are central in the context of the workplace arena; the Skills Labs are the most important to this study's focus. Apart from the institutions, much like the relationship between journals and individual experts in the public arena, individual educators can also be influential for transformation in the workplace arena. Furthermore, commodities are central for everyday work in the workplace arena. Finally, the "Bau- und Liegenschaft NRW" institution, which falls under the North Rhine-Westphalian local government's jurisdiction, occupies a unique position at the intersection of the workplace and legal arenas, because it is responsible for allocating buildings (i.e., deciding which university can use which location for a Skills Lab). While this falls under the jurisdiction of the legal arena in terms of the acquisition of new buildings, once a building has been allocated as a Skills Lab for several years, its design, interior and use all become part of the workplace arena. The border-crossing issue, therefore, is a question concerning the evolution of a given Skills Lab project.

Overall, the public arena provides the workplace arena with quality indicators. For example, associations such as the GMA and its Committee for Practical Skills issue guidelines or suggestions as to how the workplace arena might better position itself or what it should pay attention to in terms of certain quality features. Conversely, the workplace arena provides a reality check for the public arena, pointing out how certain requirements are unfeasible or are not considered relevant for everyday work.

The relationship between the workplace and the legal arena can be defined in a similar way. The legal arena establishes the formal rules and the workplace arena then adapts these formal rules according to its own judgement. The public arena, by contrast, plays a critical role by offering feedback and, in instances of uncertainty, even provides suggestions. A particularly well-known example of this phenomenon is the National Competence-based Learning Catalogue for Medicine (NKLM), where experts have collaborated to produce a quasi-legislation of their own. This can be regarded as a form of self-authorisation. It is evident that the legal arena also provides the public arena with guidelines, albeit only in the sense that it creates certain legal facts that are referred to, and thus structure, the public discussion.

Another distinctive feature is that singular experts and singular teachers can indeed be synonymous; that is to say, they can literally be the same person. Consequently, there can be designated 'border crossers' who move back and forth between two arenas (i.e., reporting on their day-to-day business as experts in the public discourse and then also bringing the public discourse to bear strongly into their work). To illustrate this, one may consider an individual who is involved in a GMA committee and who establishes standards within this public arena, while concurrently functioning as an educator within a particular Skills Lab in the workplace.

The figures and descriptions provided offer an overview of the system and stakeholders of medical education in Germany, with particular reference to SBT's methodology and its institutions of Skills Labs. However, while this considers the meso–level of institutions, and the interdependence between the professional arenas where jurisdictional claims can be voiced via different entities, the abstract knowledge of medical education also needs to be provided. This includes the structure and content of medical studies in Germany and the contested abstract knowledge of competence-based education, which substantiates the introduction of SBT as a methodology for training practical skills and competencies.

3.2 Design of German Medical Studies

Structure of the Study Programmes

German universities are formally subordinate to the federal state (Land) in which they are located. The federal states vary greatly in size, economic and social importance and the number of universities that they house. A cross-state description should, therefore, be done carefully. The difference between state and private universities is also central when describing the universities and their importance. Most German universities are state-run, but some private universities, which sometimes offer international study programmes, also exist. However, all programmes can be differentiated with regard to one major feature: whether or not they follow a regular study programme or a model study programme. The regular study programme is defined as the examination order and, therefore, sets the standard. By contrast, alternatives in so-called model study programmes are granted a reorganised curriculum while maintaining the same assessment outline and basic contents.

Thirty-seven of the medical faculties in Germany are state-run (see Appendix 1). This study considers state universities in North Rhine-Westphalia (NRW), which is one of the largest Land in terms of both population and number of universities. Eight state universities are considered in this study (Aachen, Bielefeld, Bochum, Bonn, Cologne, Essen and Münster), while one private university (Witten) is excluded. In North Rhine-Westphalia, Aachen, Bielefeld, Bochum, Düsseldorf and

Cologne follow the model programme; the remainder follow the regular programme.

In Germany, the duration of medical studies is six years, with practical skills being instructed at various points throughout the programmes. Various changes were made to the course of study during the period covered by this study, from the 1980s to the 2020s. These changes will be analysed in greater detail elsewhere (see Chapter 5). However, two exemplifications are worth mentioning. The first example is that graduates of the medical degree programmes were granted the status of doctor in internship until 2004. Consequently, the termination of their novice status as a professional was delayed for a further 18 months, until after the conclusion of their studies. This regulation was subsequently abolished, given that the title of doctor in training was associated with a lower salary, consequently resulting in a decline in the number of graduates applying for this position. Conversely, fully trained doctors are now employed directly as assistant doctors and are formally regarded as equals. However, the question of whether students acquire equivalent practical skills during their studies remains a subject of debate and is central for the debate surrounding competence-based education. The second example relates to the alteration of the undergraduate degree programme from two phases to three in 2014 (Penter & Augurzky, 2014, p. 82). Studies comprised of two sections up until 2013, with central examinations taking place after the first two years, and then again after the last four. Since 2014, however, the structure has been changed to three stages: First, there is the initial two-year foundation programme, then there is three months of basic care service and first aid training and this is concluded with a written and oral practical examination. After this stage, three more years of study and clinical placements in outpatient care follow, in a clinical environment, and finally placements in general practice are completed. The second stage concludes with written state examination, and the third is a practical year comprising internships in internal medicine, surgery, general medicine and an elective subject. The third stage is completed with an oral-practical state examination (Miani et al., 2015, p. 20).

In addition to the state examinations, described above, each subject might also entail assessments, which are organised differently in each location (Miani et al., 2015, pp. 30–31). This is due to the link to the university clinics and their respective specialties on the one hand and due to the variance of focus the educators themselves choose for their university courses and on the other.

The medical faculties and their organisation play a key role in medical education in the German educational system. While it is possible to compare different study programmes in terms of their uptake of the state examinations via the examination order, it is essential to also consider how each individual institution has unique admission processes, curricula and personnel structures. The comparison of medical departments in NRW is, therefore, advantageous because while the legislation towards the Land legislation and examination order is identical, the individual locations offer a variety in terms of both the seniority and type of study programme.

However, and beyond the formal organisational structure, the professional practice of medical education is intricately linked to the application of abstract knowledge. In the context of SBT, and in terms of the acquisition of practical skills in medical studies more broadly, the debate surrounding competence as a leading didactic concept must also be introduced. This is due to the fact that educators often ground their cognitive claims in the superiority of competence-oriented education.

Study Contents: Competence-Orientation Debate

Medical education and the academic system are connected not only through institutional affiliation with a university, but also through educational concepts and didactic theories that medical educators can offer to medical professionals. These are necessary for the training of the next generation of medical professionals (Abbott, 1988, pp. 72–73). Therefore, the concept of competency-orientation, which is connected to the professionalisation of medical education, and SBT as its output require a description.

Medical education was characterised by a dichotomy, between theoretical and practical knowledge, in the past. The difference lies in how medical knowledge is gained. Theoretical knowledge is learned from books about human biology, chemistry, physics and the like. It also covers sociological and psychological concepts, such as the patient-doctor relationship and healthcare systems. Clinical knowledge concerns the practical application and management of patients, instruments and work environments in the medical profession.

The belief was that basic knowledge had to be mastered before clinical knowledge could be gained, so traditional study programmes separated the two. However, studies using long-term progress tests showed that the two types of knowledge differ in how students learn. Students build on their basic medical knowledge over the first six semesters, but then forget some of it between the third and fourth years. Clinical knowledge is gained continuously during their studies and students reach similar test scores in both domains at the end of their studies (Görlich & Friederichs, 2021, pp. 4–5). A key argument for reforming medical education is to combine theoretical and clinical knowledge earlier in study programmes. Its proponents argue that the separation of basic and clinical knowledge can be eliminated by orienting medical education around competencies, with Simulation-Based Training being one example of this. However, educators' understanding of competence varies.

Klieme et al. (2008) provide a comprehensive description of the general educational concept of competence, exploring its relationship to education as a means of reproducing knowledge in a contextualised manner and of cultivating responsible life skills. In this regard, the educational concepts of assessment are oriented towards the outcomes of educational systems, as evidenced by the PISA study. Two international key concepts, together with other reform pedagogies which preceded the theory of competence-oriented learning, comprised the basis of McClelland's claim to shift away from classical intelligence tests and to focus on learners' practical performance in the 1970s, and Weinert's building of a general taxonomy of different competencies, defined as contextualised, meant that different competencies needed to be learnt in different practical situations (Klieme et al., 2008, pp. 7–8; McClelland, 1973; Weinert, 2001).

The adoption of this competency orientation in medical education was, therefore, advocated for; this occurred for two main reasons. Firstly, it was argued that this approach would focus on educational outcomes, thereby including learning goals and the alignment of teaching and testing (Wissenschaftsrat, 2018, p. 40). Secondly, the professional skills acquired could be as closely aligned as possible to the requirements of the professional environment in which future physicians would work, thereby making them highly specialised (Fabry, 2022, p. 288). Accordingly, reform pedagogy concepts towards competence-oriented education became influential (i.e., Miller's pyramid (Miller, 1990)). Epstein and Hundert established an influential definition of medical competencies in 2002 (Epstein & Hundert, 2002), which was followed by various international examples of medical education programmes reformed around competencies (Kamp et al., 2021, p. 122). In Germany, there are ongoing efforts to reform education towards a competence-based model, which is closely associated with SBT (see for more details Chapter 5 and 6).

Both the abstract knowledge, in the form of cognitive professional claims, and the formal organisation of medical studies in educational programmes are influential because this study focuses on a subpart of medical education and on its professionalisation in the form of Simulation-Based Training and Skills Labs as their institutional output. Additionally, it is necessary to consider what qualifies as SBT, how it originated even prior to the timeframe under consideration (since the 1980s) and which surroundings and commodities Skills Labs require to ensure the effective execution of SBT, given that it is a practical training technique.

3.3 Simulation-Based Training (SBT)

Institution: Skills Labs

The definition of Skills Labs as institutions over time must consider how SBT, as a training technique, has reproduced its practical and competence-orientation to the extent that it is represented by its own spatial and organisational format within medical faculties today; this follows the basic definitions provided by Abbott and HI, namely that institutions reproduce expertise (Abbott, 1991, p. 20) and are building blocks of social order (Streeck & Thelen, 2005, p. 9). Accordingly, it might be helpful to consider which institutions call themselves Skills Labs and what their common denominator might be. However, the first obstacle is that this study's institutional and spatial object has different wordings and spellings. These include Skills Labs, skills labs, Skills Laboratory, Simulation Centre and many others. Many authors describe slight differences in meaning; some consider a Skills Lab to encompass the entire institution in which practical skills can be taught and learnt, while Simulation Centres or Rooms are defined as spaces furnished to resemble a patient's room, a doctor's office and so on. Some talk only about simulation centres for the whole institution. My case studies also show that the sites themselves are very diverse, whether due to relocations, construction work or changes in use and output. Therefore, a very minimal definition of a Skills Lab – or synonymously a Simulation Centre – follows the functions of the spaces that represent what the process of Simulation-Based Training in its basic form should offer: A Skills Lab needs "(...) a minimum of three rooms (simulation room, control room, debriefing room)" (Jaki et al., 2018, p. 23).

This definition by Jaki et al. (2018) outlines the functions of each room and describes how SBT in its basic format must be executed to reproduce practice- and competence-oriented training. The rooms can be described as: 1) The space – which can be more or less furnished – in which acting on simulated patients or in simulators can take place. 2) A space where students or educators can observe or, depending on the design of the simulation room, control high-fidelity simulators or give further instructions to simulated patients. I have seen cases where the simulation and the control room were one room, where the simulation was in the middle of a classroom and the students sat in two rows to the left and right and watched. Here the room has no walls and this interferes with the simulation (see Chapter 5.3 for a more expansive account of this), but the separation of the room is still clear in that the roles of who is simulating and who is observing are still defined. Finally, 3) a debriefing room where the simulation witnessed can be analysed and understood as exemplary, well done, leaves room for improvement or could have taken an alternative direction and practical and theoretical knowledge can all be linked. Again, a debriefing room can theoretically be the same room as the simulation room, depending on the commodities, but often a change of space, not only of roles, offers a separation and a more open venue for reflection. What makes Simulation-Based Training interesting as a space for medical education is its practical outlet, linked to structural features (e.g., double-sided mirrors in control rooms), but even more so its commodities. These spatial arrangements and the commodities used are all part of the SBT educators' professional practice and concern their consideration for successful practical training.

Another important factor is the curricular embeddedness of these trainings, which concerns two main features. First, who is teaching within the Skills Lab? (i.e., this is not only lecturers or professors, but also student tutors in some cases, acting in the capacity of employed peer-to-peer teachers). The second features concerns whether or not the teaching in the Skills Lab is part of mandatory or facultative curricular training. Peer-to-peer learning, understood as the tutoring of students by other students, is – as seen also in this study (see Chapter 5.3.) – rather common in Skills Labs. A 2015 study found that 33 of 36 universities (excluding Bielefeld, which had not been founded yet) used this method; of these universities, 19 had at least partly administrative active tutors and three were entirely run by students for students. Thirteen programmes focused on facultative training and another 13 focused on both facultative and mandatory training. Only four focused on mandatory training (Blohm et al., 2015). The composition of personnel and the curricular embeddedness can all be considered important factors in SBT's professionalisation and recognition (or lack thereof).

In conclusion, it is evident that the configuration of a Skills Lab, both spatial and conceptual, is variable and necessitates contextualisation. Skills Labs also require a considerable commitment of finances and equipment on the part of the universities. The case studies of eight locations within this study will further illustrate the divergent circumstances in which Simulation-Based Training programmes are institutionalised and how they are, more or less, encouraged at medical departments.

Historical Origin

The origins of Skills Labs as institutions within universities are contested. This is partly because the description of a teaching space similar to a Skills Lab would not be called that today. Teaching methods or commodities were also commonly subject to historical transformation. Nevertheless, international examples, such as the University of Maastricht in the 1970s, are frequently cited as leading the way (Bugaj & Nikendei, 2016, p. 2). The following chapter provides more information about the origins of these teaching spaces and about their commodified traditions, underlining how they offer an alternative to established educational methodologies, such as theoretical lectures, book knowledge or multiple-choice questions, all while remaining part of the academic system in universities and their architecture from their earliest inception.

The oldest spatial indicators of medical practical skills training are anatomical theatres, which date back to at least the 14th century (Beese, 2024; Eckart, 2019). These theatres were used as educational showrooms for anatomy and practical skills in dissection. Students and researchers would observe an instructor performing a dissection in a room equipped with gallery seating. This procedure is similar to a modern Simulation-Based Training session, which uses plastic humanoid bodies. Furthermore, students have collected, preserved and crafted singular organs or body parts from different materials (i.e. organic tissue, wool, or wood) since at least the 19th century (Atzl et al., 2013, p. 126). These could be considered the predecessors of simple models or dummies. To that end, some argue that anatomical theatres and anatomical models should be considered as simulators. Their purpose is twofold: to represent anatomical knowledge and to convey it to apprentices of the medical craft and profession. Anatomical 'simulation' has been used as early as in Ancient China through Renaissance Italy and on into Industrialised Germany, (Owen, 2012; Bradley, 2006).

Nevertheless, contemporary Simulation-Based Training (SBT) began in the second half of the 20th century with the introduction of (plastic) simulators and simulation actors and actresses playing patients. The first simulators of this kind emerged in the 1960s in North America and Europe. One was the plastic figurine 'Resusci-Anne'. It represents the first 'cardiac compression simulation', or CPR, from the early 1960s (Jones et al., 2015, p. 57). Another example is the full-bodied 'Sim One' dummy from the late-1960s, showing symptoms of vital signs through technical enhancement. Its life-sized successor, 'Harvey', a 'cardiology patient simulator' from 1978, was developed through further testing and improvement. The development of more advanced technology in the 1980s and 1990s led to increasingly immersive simulators. One example is the 'Comprehensive Anaesthesia Simulation Environment', which reproduced a patient's anaesthesia, was almost life-like and was controlled using a Macintosh computer (Buck, 1991, p. 21; Jones et al., 2015, p. 57).

Such historicisation serves to outline that while SBT was selected for this study for its didactical concept and development first and foremost, it also entails concrete materialisations that need craftsmanship and engineering and are, therefore, closely related to the 20th century's science and technology developments, particularly in terms of plastic production and computerisation (Meikle, 1995; O'Regan, 2021). Although the educators may not consider this history in their everyday professional practice, they can nevertheless be seen as successors of the early adopters of these new educational commodities. The need for new instruction methods, not only in medicine but in general crisis and catastrophe training to, must be considered in order to understand the motivation for the early innovators (St. Pierre, 2018, p. 4). Accordingly, Bradley argues that emergency medicine and anaesthesia were naturally prone to adopting SBT, due to their emergency or non-routine nature (Bradley, 2006, p. 256). The examples of specific simulators, provided above, demonstrate progress towards the goal of creating a comprehensive environment, which gives students the opportunity to hone medial skills through experience.

The materials mentioned require further definition and follow the introduction of the Skills Lab's institutional context within the university environment and the historical background of these training programmes. As Abbott defined it, the materials of SBT are professional commodities that embody professional practices.

Commodities

SBT's commodities are their embodied knowledge. According to Abbott's definition, each commodity entails the professional practice in which it is involved. For SBT, we must understand this in a twofold manner. On the one hand, the commodities must be selected and managed by educators within the Skills Labs for the respective learning sessions and goals; on the other hand, the commodities are the learning materials with which students interact to develop specific professional skills through practical experience. SBT's commodities can be separated into three main categories: Simulators; simulated patients and commodities on the verge of SBT.

SBT's Commodities: Simulators

As outlined previously, there is a substantial body of literature that addresses the question of what might or might not be considered a simulation. The same debate extends to the materials and to SBT's commodities (Wojcik, 2022). The existing literature does not provide a consistent terminology that includes different references, languages and spellings using terms such as Puppe, Modell, Fantom, phantom, manikin, mannequin or dummy to refer to simulators. These commodities may have different functions and purposes and they can currently cost between a few hundred and several hundred thousand euros (Issleib et al., 2018, p. 62). The following three categories of simulator are proposed, and are distinguished by their respective abilities, in order to provide a more precise description in the analysis of the professional practice of SBT (see Chapter 5 and 6).

Simple models, such as the image of a 3D print of a heart, to represent only a single organ. Another example of this is a suture pad, where students learn how to sew and a representation of sections of skin is provided.

Simple dummies defined by their singular primary use, in CPR training for example. Typically, not a full body dummy is provided, but rather a section, such as a head and torso. These may or may not be computerised, so as to produce heart sounds for example. An additional example is that of an intubation simulation, which is mechanical and represents the point where fluid and air are injected, but this dummy can only be used for that specific demonstration.

High-fidelity manikins represent a complex training situation, which extends beyond the mere dummy itself. These high-fidelity simulators are equipped with sophisticated technology, including computers that display vital signs, thereby enhancing the realism of the simulation.

Simulated Patients and Hybrids Add-Ons

The historical origin of simulated patients, also known as standard patients or SP, is longer than that of the institutions of which they are a part. The Skills Lab's subprogrammes focuses on recruiting, employing and training actors and actresses for medical purposes and so simulated patients need to be included in the general educational curriculum (Thistlethwaite & Ridgway, 2006). It is difficult to define when and where simulated patients were first introduced because the lines between role-play, communication, and anamnesis training have become blurred with modern, clearly scripted SPs (Collins & Harden, 1998). Nonetheless, examples of their use can be found in literature since the 1960s (Barrows, 1968).

They are primarily used for communicational skills and, therefore, might have to showcase intense emotions that students have to react to, which needs to be handled sensitively in terms of the mental health of all parties involved (Kurtz et al., 2017; Spencer & Dales, 2006). Furthermore, simulated patients can also be combined with hybrid add-ons, which include simulators, mostly simple models, that can be used in combination with a simulated patient to ensure that students have to work with the simulator hands-on, but that they can also communicate with the simulated patient (Brown & Tortorella, 2020).

Commodities on the Verge of SBT

The following paragraphs give an overview of commodities on the verge of SBT. These include materials and embodiments for practical training, often found in Skills Lab and SBT, but which are not exclusive thereto. In some cases, students continue training outside of direct teaching (i.e., in their own studies or free training in the Skills Lab). In some cases, these commodities include specific technology and circumstances to come into effect, which is why they are used less frequently by educators.

SBT follows a specific procedure, allowing students to work on a simulator or SP. Consumables and medical instruments are often also included; a stethoscope, for example, allows students to practise sensory skills (van Drie, 2013). Simulators and SPs are designed to help students prepare for professional practice. Medical gloves, for instance, establish a sterile environment and teach students about hygienic practice (Underman, 2022, p. 188). These instruments and consumables must be introduced, applied and managed by educators and students, even though they are not designed or used in SBT exclusively. (Virtual) patient vignettes are realistic case descriptions that can help students to learn. They use diverse media, like explanatory text, pictures, sounds and digital formats to represent the digital version of a regular study book (Huwendiek & Haag, 2014). They can be understood as being on the verge of SBT because they have a considerable overlap with the 'modelling' of patients' cases with high-fidelity manikins. These digital programmes showcase patients' data and control a high-fidelity manikin; both also have the same concept of patients as data bundles. Furthermore, virtual patient vignettes stem from similar advances in technology as SBT, due to their computerised nature, while also originating in the reform pedagogy that asks for problem-based, self-sufficient learning environments. Patient vignettes are often also meant for self-study by students, while SBT, as mentioned previously, is largely reliant on the shared space that houses the commodities in Skills Labs.Virtual Reality (VR) is used in medical education for 3D computer simulations. Students observe a 3D-modelled structure, such as a heart, from multiple perspectives, including different layers of tissue, to understand its anatomical features. This experience can be made more immersive with VR controllers and with hand motions. However, all parts are digitally dematerialised. Thus, students see and interact with 3D objects in a digital space, without holding an actual educational commodity (Bowyer et al., 2008, p. 91). In comparison to VR, Augmented Reality (AR) includes a commodity. In AR, the same 3D models and glasses are used, but these are projected onto real objects with which students can interact haptically.

Another form of virtual reality with augmented features is the immersive projection room in which videos are projected onto walls. These are used for team training and for extreme emergency situations, such as natural disasters or war, where aid is required quickly and where emotional support is required (Bowyer et al., 2008; Ilgen et al., 2013; Wier et al., 2017). VR, AR and immersive projection rooms are similar, because of their digital nature, and are comparable to high-fidelity manikins because both represent the body (partly) digitally.In consideration of the experimental origins of Skills Labs and their commodities, a notable example is the utilisation of self-made materials, like thread and embroidery to represent tissue (Kneebone & Oakes, 2022), and fruit and textiles to simulate a uterine inspection and procedure (Shea & Rovera, 2015). This approach involves sensory learning with everyday objects and is unlike classical SBT, which employs pre-produced objects such as high-fidelity simulators. In conclusion, a thorough analysis of Skills Labs is required to determine its institutional and organisational structure within the medical educational system and in order to assess its subordinate role within the medical profession. SBT is founded upon academic knowledge of didactic concepts, with a particular emphasis being placed upon competence-orientation. The spatial and commodified context indicates the long-standing tradition of objectified learning in medicine, as well as the introduction of complex technology. The construction and implementation of these commodities bestows an expert status on SBT educators. However, this should not obscure the analysis of the juxtaposition of the 'old' and 'new' commodities being used and the embodied knowledge that they represent (Edgerton, 2008, p. xvii). These elements collectively signify an understanding of learning as a bodily experience, fostering the development of professional expertise through training and the mastery of knowledge and practical skills.

3.4 Summary

In conclusion, as a methodology of medical education, SBT is connected to diverse entities and stakeholders that need to be considered in order to analyse and to understand its rise at German universities since the 1980s. This is best achieved by considering three professional arenas and their logic of operation. In the context of the legal arena, the primary concern of analysis must be legislation, due to its standard setting of content and structure of medical education. This analysis must also consider the difference between the nation-wide binding examination order, issued by the Federal Ministry of Health, and the individual university curricula, which can expand the former towards their own focal points. The public arena, however, is best considered in relation to the debates surrounding standards in both the legal and workplace contexts, frequently critiquing the former and proposing improvements to the latter. Statements of association and arguments presented within journal articles, which are part of the academic system, represent central materials for the analysis of positioning and about the regime in which it is deputised. Lastly, the workplace arena includes individual Skills Labs and the composition of SBT. Requirements of medical studies, according to the examination order and didactic concepts aimed at competence-orientation, must be considered in order to understand how educators interpret the design of the Medical Educational System (MES), and attempt to adapt it to their advantage. Ultimately, SBT is best understood in terms of its transformative nature, as evidenced by its institutional output of Skills Labs and by the allocation of space and commodities. SBT's transformation further illustrates the close relationship between technological development and the rise of SBT. The diversity of SBT's contemporary commodities once again emphasises its practical, experience-oriented nature.

A multi-perspective analysis is also required in order to achieve a comprehensive understanding of these aspects, and of the three professional arenas for the professionalisation under investigation in particular. The following chapter outlines the data that was utilised and the methodological procedures that were employed, which is referred to as a 'mixed-data approach'.

4. Study Design: A Mixed-Data Approach

The meso–level of social interaction is central for the given study and in terms of the social and institutional transformations examined. Cognitive and social claims need to be considered with respect to the three jurisdictional arenas (legal, public, workplace) and following Abbott's argument towards jurisdictions. HI extends these considerations towards a (historical) process description for a larger timeframe than is often considered in sociological studies. This brings about the need for time-overarching, triangulated data that sufficiently allows for the analysis of both the formal and informal composition of institutional and professional circumstances. Two frameworks are used to achieve these goals: complex reasoning, circumventing traditional problems of validity, reliability and generalisability. A mixed-data approach has been deployed that follows the reasoning of mixed methods approaches. This chapter describes the methodology employed to analyse SBT's introduction and transformation in Germany in two subchapters: first, a brief, general discussion of the methodology employed in this study is provided followed by a description of the research procedure, outlining the materials and analysis.

4.1 Methodology

General Quality Criteria of Social Studies

It is crucial to consider the quality criteria that inform the selection of research methodologies in the context of social studies. These criteria illuminate the different levels at which social phenomena are studied as examples for transformation, ranging from micro- to meso- to macro-levels (Krause, 2021, p. 6), or in terms of which methodological schools are considered to be influential (Otte et al., 2023), or which quality criteria can be considered to be practicable (Kelle & Reith, 2023, p. 18). Consequently, a variety of approaches and methodologies provide different opportunities for analysis. The present study has elected to adopt a qualitative, mixed-data approach, orienting itself towards a mixed-method approach. It is a commonly accepted principle that the validation criteria employed in the field of social studies include generalisability, validity and reliability. The following criteria can be expressed in concrete steps for the purposes of qualitative analysis, as described by Maxwell (2013) and Lareau (2021).

Maxwell provides a list of eight checkmarks for the validity of qualitative studies, while Lareau provides four answers to the so-called 'four Rs'. Maxwell's list aims to bridge the gap between the (seemingly arbitrary) choices in data analysis in qualitative research and the goal of validity. Intensive and long-term involvement (I) can help to ensure that the observed social situation is not an outlier and produces rich data (II), in different observation points, possibly by different observers, or the expressiveness of in-depth interviews, etc., with the possibility of returning to the community being studied for their respondent-validation (III). This, in turn, requires a reflection on the interventions of qualitative research (IV), which means that the observation of a social situation can have an effect and that this must be a part of the analysis itself. The inclusion of a variety of cases, like the inclusion of discrepant evidence and negative cases (V) for the situation being studied, helps to ameliorate the power of these threats to validity. Moreover, the triangulation of data (i.e., the inclusion of historical documents, the conduct of interviews and statistical background information) helps to enrich the data set, as does the triangulation of analytical approaches, also known as mixed methods (VI). Finally, numerical information (VII) can support qualitative claims and can also help to draw comparisons by matching similar characteristics (VIII) (Maxwell, 2013, 126–129). Although this list can be criticised for not being exhaustive, and because not all studies can include all eight criteria at the same time, Maxwell's list remains useful for broadening the understanding of the possibilities that qualitative data offers for analysis, without even talking about the epidemiological background assumptions offered by certain methodologies and by their theoretical backgrounds.

Another strategy is to respond to the so-called "four Rs", which concern the main criticisms of qualitative studies proffered by quantitative researchers, according to Annette Lareau (Lareau, 2021). Firstly, *representativeness*, which is the aim of making data generalise to a whole population, also called generalisability. Lareau's answer is that qualitative research 'goes deeper, not wider' with its findings. This refers to the level of research, in this case the micro- rather than the macro- level of analysis. It, therefore, offers a different perspective on social phenomena. Secondly, *reactivity* is the problem of participating in the social situation by observing it. Not only does this apply to quantitative methods through interviewer effects and biases in quantitative procedures but Lareau's argument also points to the exhaustion of people to act differently over a long period of time. This is directly related to the first, second, fifth and eighth points in Maxwell's list, because long-term involvement and rich data with multiple field entries leads to greater trust with participants and more observational assessment of a situation's naturalness, so that people come to forget that a researcher is present. While these effects cannot be completely eliminated by the very nature of observation, the search for either contrasting or characteristically similar cases helps to facilitate comparisons between data. Thirdly, *reliability* is best addressed by the triangulation of data, in Lareau's reasoning, which is also consistent with Maxwell's view. Finally, *replicability* in quantitative procedures suggests that data should not be tied to the researcher who conducts it and should, therefore, be reproduceable at any time by any person. The problem of replicability in qualitative research is potentially twofold: Lareau speaks of the issue of gatekeeping and the aforementioned trust necessary to conducting interviews, making observations and sometimes even the analysis of sensitive documents. Her argument is that any researcher who overcomes gatekeeping should be able to produce the same results as the qualitative study in question (Lareau, 2021, pp. 34–35).

Indeed, the process of looking for characteristics between cases, then diving into the data and extracting comparisons, which can then be reflected back to the overarching research question, is a good example of an effective strategy for argumentation in (qualitative) social research. Creswell calls this "complex reasoning" (Creswell, 2013, p. 45). This description of reasoning does not just concern analysis, but is also already built into data collection and reflects flexibility. The iterative process involves organising the understanding of findings into categories (inductively) and then theorising the categories further and confronting them with the data (deductively). In summary, *generalisability*, *reliability*, and *validity* – or in other words representativeness, reactivity, reliability, and replicability – are useful criteria for assessing whether a social study has been conducted well and if methodological threats have been avoided. Furthermore, complex reasoning can also be considered a component of the mixed methods methodology (Kelle & Reith, 2023). The present study can be understood as mixed methods in a rather broad sense and one could characterise it as adopting a mixed-data approach with a strong focus on triangulation. Some of the methodological arguments in the mixed methods literature are briefly outlined in order to underline this argument, before clarifying how triangulation was achieved in greater detail.

Mixed Methods and Triangulation of Data

Mixed methods are a third way of looking at methodology in the social sciences that has been gaining ground since the 1990s. However, there is no single definition of what mixed methods might be. In general, mixed methods by their very name call for a blending of the traditional qualitative and quantitative camps within one study. The advantage of this approach lies in the different levels of social phenomena that can be studied (Knappertsbusch et al., 2023). At first glance, this approach may require entirely new methods. However, epistemological assumptions associated with the different types of analysis and data collection can be strong, which may make it inappropriate in the eyes of critics to mix them (Patton, 2015, p. 656). For this reason, some proponents of mixed methods view it more in terms of a connection of different types of methods, qualitative and quantitative, in a single study, investigating the same research question. Specifically, this means either starting with a quantitative dataset and exploring interesting patterns in a qualitative way or starting with an exploratory qualitative dataset and deciding which parts are worth investigating further quantitatively from there. This procedure requires a very goal-oriented sampling of data, in a so-called simple, clustered or stratified way, so that whatever sequence of analysis is chosen can be performed on the given data set (Tashakkori & Teddlie, 2009, pp. 290–291). Other authors echo this stance of not opposing quantitative and qualitative methods, instead using both in a pragmatic way, all while questioning whether this can be called mixed methods or rather a bricolage of different approaches. This refers to the fact that the different types of data and analysis still retain their inherent logic and that they are not affected by 'mixing' (Gobo, 2023). Others still argue that these procedures may not even be called mixed methods, but that they should simply be described in terms of the procedures followed and, where possible, should use different types of data (Symonds & Gorard, 2010, p. 133).

Triangulation generally follows the same logic originating in land measurement: "The term 'triangulation' derives from surveying. Knowing a single landmark only locates one somewhere along a line in a direction from the landmark, whereas with two landmarks one can take bearings on both and locate oneself at their intersection" (Fielding & Fielding, 1986, p. 23). Jane and Nigel Fielding describe in "Linking data" (Fielding & Fielding, 1986) how researchers can compare cases across different aspects, without expressing the idea of 'hard' quantitative facts or 'soft' qualitative descriptions, but by using any data that helps to strengthen comparability (Fielding & Fielding, 1986, p. 10). Within this rationale they acknowledge the same 'problem' as in the discussion of mixed methods outlined above, that different methodologies have different inherent logics. However, they point to the numerical part of qualitative surveys by at least counting their cases, and the descriptive parts of quantitative methods, firstly through the questionnaire administered by the researchers (potentially revealing their biases), and secondly in the underlying assumptions of probabilistic laws of the social that make the interpretation of statistical models feasible (Fielding & Fielding, 1986, p. 12). Their argument for triangulation concerns the importance of mixing data to improve the quality of any social study. In practice, this means using different sources for cases, varying positions within a discourse, utilising different points in time for a historical genealogy, and possibly even including other researchers or the feedback loop of the scientific field to improve the evidence supporting the claims made by the study (Fielding & Fielding, 1986, p. 25). This comparative method rationale is tied to the social situation, given that it includes not only the micro–sociological interaction, but also the underlying context of social laws, which are inevitably intertwined at the micro- and macro- levels, as Fielding and Fielding argue. This not only involves moving from the micro- to the macro- level of explanation, but rather mixing or circulating the inner logic of "analytic induction" to "hypothetico-deductive", back and forth repeatedly (Fielding & Fielding, 1986, pp. 14–18). Again, this method resembles complex reasoning, as defined by Creswell previously.

Creswell himself also illustrates this with the example of intercoder testing, a procedure whereby documents are analysed using different methods and the results, called codes or markers, identified as important to the research question in the given text are 'tested' by a second person. In this way, one person's bias about the analysis and the possibility of overlooking or overcomplicating the analysis can be reduced (Creswell, 2013, p. 253). However, intercoder testing points to the nature of social research itself, as mentioned previously. The researcher cannot be completely eliminated as a factor in the collection and analysis of data, but he or she can be made an indicator for results by being as reflective and open as possible about the train of thought and about the understanding ("Verstehen", as Weber defined it) of social behaviour, which is achieved by reflecting on what is actually observed and in terms of the social scientist's interpretation (Fielding & Fielding, 1986, pp. 35–41). Against this bias, triangulation assists: "Triangulation – collecting information from a diverse range of individuals and settings, using a variety of methods – [...] reduc[es, J.D.] the risk of chance associations and of systematic biases due to specific method, and allows a better assessment of the generality of the explanations that one develops" (Maxwell, 2013, p. 128). In general, finding inconsistencies and contradictions might not be an indication of a bad study, but is instead an opportunity to develop the validity of the results (Patton, 2015, p. 661). One way of doing this, as mentioned previously, is to trace the historical developments that have led up to the phenomena encountered in the present.

When a qualitative study addresses a historical dimension, either by arguing for a certain genesis, or by referring to certain historical events or by actually using historical data and documents, there are certain imperatives from historical research itself that should be adhered to. Hines (2011), a historian herself, advises that social scientists be aware of the following problems in historical research. In her article, she compares historians to investigators or detectives, as seen in everyday life on television or in other stories, looking for clues and trying to understand the circumstances behind an event. This can also be said to be true for a social scientist; however, where any scientist and the fictive investigator must part ways is in the 'closing of cases', because for the latter it marks the end of the show. Instead, both historians and social scientists must remain open to alternative narratives, as mentioned above in relation to the triangulation of data (Hines, 2011, pp. 138–139).

This advocacy is not surprising, given the ambiguity of the source material and the fragmented knowledge available to historical analysis, depending on the period. Examining the context of an event and understanding its wider implications from there is akin to the approach adopted by a social study. Moreover, the concept of agency is an important and contested object of historical and social analysis in a similar way. This is because attributing power to actors must be done with caution and only when there is a clear indication of this in the given materials (Hines, 2011, p. 148). Materials and data are, therefore, perhaps the biggest difference between historical and social studies, because historical sources must be divided into primary and secondary sources. Primary sources are those that were created during the period being examined, while secondary sources are interpretations of events made thereafter. There may be contradictions or counter-evidence even within a given period, so it is important in historical analysis to consider the wider context and possible conflicts in the material production, and to look for cross-links as a form of triangulation (Hines, 2011, p. 150). Materials suitable for historical analysis can, therefore, include public records or publicly published documents, such as the unchanged legal documents or scientific articles in this study (Hines, 2011, pp. 151–153). Having discussed the benefits of a historical perspective for social research, and the potential for the further triangulation and enrichment of interpretative possibilities, the present study's design will be introduced in the following section.

4.2 Procedure of the Present Study

Data Gathering and Materials of Cases

The cases can be understood on two levels for the present study: On the one hand, our interest lies in the development of a specific profession, namely the medical profession, and its educational subgroup, of which Simulation-Based Training is an example; on the other hand, the object of the study is simulation as a medical training technique, and the Skills Labs or simulation training centres as institutions that provide Simulation-Based Training, provide the different cases (i.e., the different sites). The following discussion of data gathering must be understood as occurring on these two levels, which congruently represent the meso–level of social phenomenon with SBT's formal institutional embeddedness (data type I), the broader discourse on medical didactics (data type II) and the micro- level of social organisation within the Skills Labs (data type III).

Accordingly, the three types of data used in this study highlight different parts of the transformation of medical education exemplified by SBT and the case studies of Skills Labs examined, thereby offering thorough triangulation and analysis of the contemporary state extended by historical developments.

Data Type I: Legal

The first type of data, which was chosen in line with the legal arena (Abbott, 1988), is legal documentation (data type I), which consists of two types of documents: the examination order(s) and university curricula (see Appendix 2). Legal documentation concerns the granting of the medical degree, which can be obtained at a public or private university in Germany and which offer regular or model study programmes and concludes with the obligatory state examination (Staatsexamen). Therefore, only the thirty-seven public universities in Germany and their medical faculties and study hospitals, which grant the state examination, were chosen for this study.[1] Furthermore, with a timeframe from 1990 to 2022, the starting point was chosen due to the reunification of the two German countries in 1989/1990 and the subsequent recognition of degrees and the alignment of examination regulations from West and East Germany, which was not implemented for all professions.

The first legal documents, the examination order and its amendments (Approbationsordnung) are issued by the Federal Ministry of Health and represent the nationally binding composition of the state examination with its two, later three mayor exams, the First, Second and Third State Examination (in German Erstes, Zweites und Drittes Staatsexamen, see Chapter 3.1). The Federal Ministry of Justice offers an

1 Appendix 1 offers a comprehensive list of the locations included in this study and the following institutions were excluded: Oldenburg's university programme, due to their cooperation with the university of Groningen, Netherlands, which represents a unique, international structure. Private faculties, such as Paracelsus Priv.-Uni. Nuremberg, Priv.-Uni. Witten-Herdecke, School for Medicine Kassel, Asklepios Medical School Hamburg or Medical College Brandenburg were also excluded.

online archive of past and present laws, regulations and amendments, called Bundesgesetzblätter, from which twenty-one examination orders in their different versions could be obtained for the years 1990 to 2022, and one from 1979, for the purposes of comparison.

A lexical search was also conducted using the search terms "skill*" and "sim*". The asterisk indicates all possible combinations of words within or between words; for example, the search included skills, Skills Lab, skills training, and so on, or simulation, simulation patients, simulation training, and so on. The following step involved using Adobe Acrobat's side-by-side view for the purposes of comparing papers that were published in close conjunction with one another. All the documents were subsequently transferred to MAXQDA, a programme primarily designed for qualitative social analysis. This process enabled the execution of a directed content analysis that entailed the application of fine coding. The second batch of legal documents concerned the curricula of all thirty-seven universities within the sixteen German Länder that have state-run medical training programmes (excluding dental, veterinary and specialist programmes) between 1990 and 2022. They were collected via the homepages of these programmes, university archives or via requests made to examination offices. This resulted in 341 documents of interest and were fetched through several rounds of requests.

Data type I clarifies the relationship between the national examination system and the university programmes associated with the sixteen Länder. While the development and administration of the examinations must be uniform across all locations, the actual curricula and teaching techniques can vary. It is evident that not all thirty-seven university curricula explicitly mention either simulations or simulation centres as an integral component of the teaching curriculum. However, it is notable that all of these curricula feature a facility of this nature on their official homepages, thereby underscoring a discernible distinction between broader discourse and legislative frameworks. In consideration of the aforementioned observations, it is reasonable to examine the case of one federal state, namely North Rhine-Westphalia (NRW), which is home to eight universities that offer such programmes; of these, three are regular study programmes and five are model study programmes, with all programmes concluding with the same state examination. In addition, all eight universities were equipped with a Skills Lab, though the implementation periods, teaching techniques and facultative versus mandatory curricula all vary. These techniques include peer-to-peer teaching and expert teaching. All of the above offer interesting points of comparison between locations.

Data type II: Public

The second type of data used consists of scientific articles (data type II) taken from a specific journal, the Zeitschrift für Medizinische Ausbildung or the Journal for Medical Education (JME), published in English since 2016, which has been steadily gaining ground since its inception in 1984. This journal is important because it is published and edited by the largest association for medical education in three German-speaking countries (Germany, Switzerland and Austria), namely the Association for Medical Education (GMA). A pre-selection of articles was made from the journal's online archive, with a focus on Germany from 1984 to 2022, which covers the journal's introductory volume (i.e., its founding issue and self-description); the transformation of medical education examination (i.e. the consideration of different types of learning or examination systems, the discourse on competence, etc.); the rise of Simulation-Based Training (i.e. the use of simulation patients, which is mentioned as early as 1984) and the editorial, mentions of conferences, new books or other celebrations. Papers on dental medicine or veterinary medicine or other special issues were excluded. This resulted in a total of 1,419 articles from 125 issues, which were further processed by a lexical search for a content analysis using predefined keywords for a summative content analysis (the procedure for the use of MAXQDA is described below).

The formation of 'sets' was conducted on the basis of keywords that were deemed to be highly conducive to the research endeavour. The set of relevant documents, after the summative content analysis, comprised 499 documents that were finely coded, in addition to a number of random samples, particularly from the journal's early days. Consequently, 527 documents were manually coded using a circular inductive-deductive scheme (see Appendix 3). Despite meticulous pre-selection, a number of articles were excluded from this set due to the presence of foreign references or erroneous coding, for example, "simul*" also occurs in "simultaneously", resulting in a total number of 448 articles that contained actually relevant content. These journal articles provide a glimpse into the public arena, as defined by Abbott (1988) and represents both the transformation of professional standards for medical education and the requirements that a well-educated medical practitioner sis expected to meet. The precise procedure established for the processing of the data is summarised in Appendix 7; they are quoted by their year, value, issue and article number to mark them as primary sources, similar to legal papers. The Appendix 3 gives an overview of all of the articles analysed.

Data type III: Workplace

For the third type of data, micro-two different types of inquiry are necessary when looking at the micro–interaction of actors (data type III). The first type is participatory observation of Simulation-Based Training in the Skills Labs at the eight sites selected in NRW in order to analyse the workplace of educators within the Skills Labs. The study's objective was to visit all eight sites in NRW between April and July 2024, which was achieved with the exception of Düsseldorf due to the construction and relocation of the Skills Lab during this period. The study's primary focus was on the analysis of the Skills Labs as locations themselves, and the observation of training (where feasible). The rooms that housed the Skills Labs, including the storage spaces and personnel offices, were visited throughout the course of a series of six appointments. In addition, a total of four training days were observed at the Skills Labs. Three training sessions incorporated SPs and two training sessions utilised simulators during these observations. In one particular instance, an observation of an examination, namely an Objective Structured Clinical Examinations (OSCE), was conducted.

The second type of inquiry consists of in-depth expert interviews (Meuser & Nagel, 2009; Pfadenhauer, 2009), extending the observations made in the Skills Labs, through oral, mostly in-person interviews with the staff of the Skills Labs selected, to gain insights into their implementation of Simulation-Based Training (see Appendix 4 for the interview guide).[2] This type of researcher-generated material (type III) is often referred to as fieldwork and is more closely associated with the identity of the researcher than it is with pre-existing documents. In addition to the various threats posed by this type of research, part of the interviewer effect may be the researcher's own connection to the field in question. As Pfadenhauer (Pfadenhauer, 2009) asserts, being both an insider (i.e., someone who comes from the field being researched) and an outsider (i.e., someone who does not come from the field and therefore does not know parts of it) can be either advantageous or disadvantageous. The phenomenon can be explained through the concept of social proximity, which states that individuals are more likely to form connections with those who share similar educational backgrounds, institutions attended, and experiences of discrimination. Alternatively, this tendency may arise from perceiving oneself as naive, leading others to assume a position of authority and to provide explanations about the topic in question. The extent to which these factors influence the outcome is contingent upon the specific research questions and on the nature of the interaction partners (Lareau, 2021, pp. 27–28). In this particular instance, the researcher has neither a medical nor a didactic background; as a result, they are often perceived as an outsider. Nevertheless, the insights generated from data types I and II prior to the interviews were instrumental in establishing the status of the interviewer as an expert on SBT organisation and institutionalisation. Furthermore, stressing confidentiality is imperative in order to ensure that interviewees feel comfortable when talking to an interviewer, even about sensitive topics (i.e., power struggles at their workplace). This involves the use of pseudonyms and the disclosure of only general circumstantial information that cannot be used to identify individuals (Lareau, 2021, pp. 38–41). It is also important to be as precise as possible about the study's purpose

2 This study's author has German as a mother-tongue, but the given study is presented in English, making translation and the parenthetical possible shortcomings in social and cultural meaning, inevitable (Mol, 2014;W.-M. Roth ,2013; Schittenhelm, 2025). Therefore, quotes are given in translated version, plus the original German wording, in some cases.

with participants, without pre-empting the views of the interviewees or the possible outcomes of the study (Lareau, 2021, p. 47).

Figure 3: Overview of interviews

Time period: 01.07.2022-31.07.2024	
∑ 16:28:27; Ø 01:01:47	
Token	**Length**
SkL01	00:55:39
SkL02	01:17:04
SkL03	00:53:03
SkL04	00:24:46
SkL05	01:00:07
SkL06	00:57:53
SkL07	01:13:39
SkL08	00:55:30
SkL09.1 + SkL9.2	01:22:08
SkL10	00:28:14
SkL11	00:17:59
SkL12	01:19:02
SkL13	00:45:10
SkL14	01:24:13

Figure 3 provides a synopsis of the expert interviews that were conducted. A total of sixteen interviews were conducted with eighteen individuals in the period between 2022 and 2024, yielding a total of 16 hours and 28 minutes of content, with an average duration of 1 hour and 1 minute per interview. It was possible to access all locations and all status groups within the Skills Labs. This took place either through recommendations from colleagues at a given Skills Lab, or spontaneously during a visit to a Skills Lab.

All interviews were transcribed and analysed using a directed content analysis. The indication of "position in text" is relative to the analysis conducted using the MAXQDA software. It is noteworthy that all interviews were conducted in German, thereby necessitating a translation by the author which was as close as possible to the original. All pronouns were modified to align with a gender-neutral form in order to enhance the anonymisation of participants.

Data Analysis: Content Analysis

After having depicted the types of material used in the given study, the second step of defining the analysis of the data follows naturally. Studies commonly tend to describe either data gathering or data analysis in longer paragraphs, but both are fragile parts of the research process that require great consideration. The data collection itself is not yet scientifically valuable until it is both evaluated and analysed in a coherent fashion, while it also defines what kind of data will be available for later analysis.

Several types of analysis would have been possible from this study's data. I will not describe alternative analysis tools, only the one that was chosen: content analysis. Content analysis has different versions and a broad methodological discourse surrounds it, especially in the German discourse (Kühlmeyer et al., 2020). However, looking at the Anglo-American methodological literature, which is more relevant here, one might define it as follows: A content analysis is bound to *text* as its core medium in order to both count and understand different kinds of knowledge within the given texts (Hsieh & Shannon, 2005; Schneijderberg et al., 2022, p. 30). The *knowledge of symbols* is given within the language, which one must be able to understand in its complex variations. In my case, having German material and being a German native speaker is valuable. Another example could be the analysis of internet culture and communication. Here the use of emoticons or smileys and their different meanings could represent a knowledge of symbols that is necessary for successful analysis. Furthermore, it is necessary to focus on the *knowledge of context* because different kinds of texts contain different kinds of information andthe historical context of a piece of text from a certain period is especially important to take into account. This might sound like a trivial remark, but in the context of the phenomenon under study here, for instance, the development of Simulation-Based Training has only been possible alongside the progress of computerisation and technology, thereby reflecting a certain historical context. Finally, every text contains the author's motivation and, therefore, the analysis should testify to the *knowledge of motivation* (i.e., the legal texts provided represent the institutional analytical point of view, whereas the observational notes represent the professional micro–interactions and so on). The important indicators here include who is addressed by the text, why they are addressed and by whom they are addressed. The combination of these three types of knowledge – symbols, context and motivation – constitutes the content of a text (Schneijderberg et al., 2022, p. 30).

The actual steps of a content analysis can be different and depend on the purpose of the analysis. Schneijderberg et al.'s workbook provides a good overview of what they identify as qualitative and quantitative steps within a content analysis. They describe some of them also as either "(semi-) automatic" or "supported by programmes", and all of them as less epistemological oriented (Schneijderberg et al., 2022, 15). This means that, on the one hand, it can be useful to use a programme to organise, rearrange and record results from the data. In this study, this is done through a coding scheme in MAXQDA, which has options for many different types of, mostly qualitative, text analysis; on the other hand, the less epistemological account that Schneijderberg et al. present of content analysis is in line with the pragmatic stance adopted towards the methodology expressed in the previous section.

In addition, Hsieh and Shannon's article (Hsieh & Shannon, 2005) presents another way of describing different, concrete content analysis procedures by dividing them into three types. Firstly, *Conventional content analysis* sticks close to the text and builds codes, meaning the practice of marking words, phrases or paragraphs of the text that are important to the research question and naming them according to their analytical meaning; it is also mostly inductive. This defines a more descriptive style of looking at the text and reporting on its content. The second type is called *Directed content analysis* and is closer to my process. Here, codes are produced not only by analysing the texts at hand, but also by theory-driven features. There is both an inductive and a deductive part and this allows codes to be produced during the analysis, as well as allowing for the 'testing' of both codes and the concept they represent by looking for them in the texts. The third and final type of content analysis, which was followed partly within this study, is *Summative content analysis*. In this variance, keywords that paraphrased theoretical or practical content related to the research object are identified and counted in order to understand a given issue's size (Hsieh & Shannon, 2005, p. 1,286). Figure 4 provides another overview of the three types of content analysis, after Hsieh and Shannon (2005).

Figure 4: Three Approaches to Content Analysis (Hsieh & Shannon, 2005, p. 1286)

TABLE 4: Major Coding Differences Among Three Approaches to Content Analysis

Type of Content Analysis	*Study Starts With*	*Timing of Defining Codes or Keywords*	*Source of Codes or Keywords*
Conventional content analysis	Observation	Codes are defined during data analysis	Codes are derived from data
Directed content analysis	Theory	Codes are defined before and during data analysis	Codes are derived from theory or relevant research findings
Summative content analysis	Keywords	Keywords are identified before and during data analysis	Keywords are derived from interest of researchers or review of literature

Two examples of detailed content analysis steps are presented in the following discussion that were carried out with data type II. These examples, drawn from the given materials, illustrate the different parts of this qualitative procedure. Firstly, a lexical search using MAXQDA will be described, which can be considered to be summative content analysis, due to the use of predefined keywords. However, it should be noted that this was merely the initial phase of the research process. Consequently, the subsequent analysis can be considered to be directed content analysis. In this manner, the research process's second stage involved the coding of all available documents relating to the research object, namely Simulation-Based Training and the development of the didactic discourse that defines medical (educational) professionalism.

The initial portion of the investigation proceeds as follows: It is necessary to ensure that computer programmes or scripts are able to 'read' the text in order to prepare documents for lexical searching, such as that facilitated by the JME articles. In this particular instance, Adobe Acrobat Pro's 'Optical Character Recognition' (OCR) functionality was utilised to convert all of the PDF documents into a more legible text for programmes and for their text searches. The documents were subsequently imported into MAXQDA, where the "Advanced Text Search" function located within the "Analysis" tab was utilised. Within this section, the user is able to enter search terms and to specify the manner in which they should be searched. This can be done in conjunction, in isolation, or not at all. The search can be conducted within a sentence, a line, or in a paragraph of text. Once the search has been initiated, the user can view the numerical output in the results, as well as the auto-coding of those results. It is imperative to acknowledge that the search terms are pre-determined by the researcher. Consequently, it is essential to be proficient at conducting an effective query. Schneijderberg et al. also include this procedure in their workbook, referring to Lemma as the best practice for searching for word stems and using asterisks (Schneijderberg et al., 2022, pp. 160–165). In the present study's context, the objective was to reduce the number of JME articles in the sample for the purpose of conducting a directed content analysis (i.e., coding only relevant documents). The lexical search was employed to identify articles containing keywords of interest (i.e., with the lemmas "simul*"). Consequently, any word containing the term "simul" will be included in the search. The documents are in German, so "simulieren", "Simulation", "Simulationszentrum" and so on are included. For example, my first sample included 79 JME articles for the years 1984 to 1989. A more detailed directed content analysis can be undertaken within this reduced number of articles. Interestingly, Schneijderberg et al. understand these kinds of word searches as quantitative, in the sense that these counts are manifest values given within the texts (Schneijderberg et al., 2022, pp. 170–171).

The main procedure, which is undertaken for all three data types, involves a directed content analysis according to Hsieh and Shannon (2005). This process entails a twofold interest in coding: On the one hand, theoretical assumptions are inherent in the formulation of the research question; on the other hand, the researcher is open to other (potentially important) phenomena and to the language of the given content in particular (Hsieh & Shannon, 2005, p. 1286). Consequently, the interpretation of symbols, context and motivation is only partially clear before the line-by-line coding is executed. In the context of MAXQDA, the screen is divided into multiple windows that each represent distinct analytical components. These components include a window dedicated to the analysis of the text currently under consideration, another that displays the list of documents that have been imported into the project and a third window that presents the list of codes, also referred to as the "codesystem" or codebook. Additionally, if the "segments of crossover of codes" function has been activated, then a fourth window will appear that refers to the "place" in which a given code was utilised. The codesystem is a method of organising significant concepts that have been identified within the text. It can be used to enumerate both the absolute number of documents and passages in the text. Furthermore, when documents are accompanied by years of origin, then the code system can provide a historical development of transformation. MAXQDA is a software that facilitates further analysis in the form of cross-tables, diagrams and other representations of the given codes. However, its primary function is to assist researchers in navigating a substantial corpus of text. Complex reasoning can be achieved through the function of memos in specific, which are connected to either a code or document (Schneijderberg et al., 2022, pp. 160–165). In conclusion, it is evident that the analytical tools of cross tables, smaller numerical comparisons, visual tools and the rearrangement of codes and subcodes are subject to continuous enhancement through further analysis and to the importation of additional documents into MAXQDA. It is evident that legal documentation is characterised by a different layout compared to scientific articles or transcripts of interviews. However, it should be noted that each type of data illuminates a distinct aspect of the research question within its specific codes. It is evident that this approach facilitates both the triangulation of data and the generation of analytical insights, thereby enhancing the robustness of the study in question.

One critique that can be levelled at content analysis, or indeed any procedure in which the researcher at least partly produces codes inductively, is the potential arbitrariness of code production. It is imperative to employ intercoder testing in order to enhance the reliability of the utilisation of codes. Creswell (Creswell, 2013, p. 253) advises that a second researcher should code material to allow for a productive discourse on the meaning of codes. This procedure was conducted from the study's early stages. The second coder was required to collaborate with the initial coder, utilising the existing codebook to ensure an understanding of the initial codebook, in order to guarantee the initial codebook's reliability. Subsequently, the second coder was tasked with coding relatively simple material, employing the existing codes as a foundation and with generating their own codes. This process involved the meticulous examination of discrepancies in either understanding or interpretation. This approach aligns with the findings of Kuckartz and Rädinger (Kuckartz & Rädiker, 2020, p. 48), who advocate for the utilisation of a student assistant in circumstances in which a team is not involved in the research process, a condition that was met in this instance. The result is that memos are immensely helpful, not only in explaining the meaning of certain codes, but also in making the second coder self-sufficient in coding. New aspects within the given material, which were not part of the initial coding, were identified and added to the codebook and larger contexts were seen and understood in the same way as the first coder understood them. Other codes were questioned, concerning their necessity for inclusion, but were better understood in the debriefing.

A further critique of the present study is that it is qualitative in nature and, consequently, that it is only able to examine smaller components of the phenomena under investigation. The argument has been made that a pragmatic stance might be more beneficial in such cases, with the triangulation of data being identified as a priority for any study wishing to achieve sufficient generalisability, reliability and validity. In this instance, a representation of the entire population of relevant universities is available: A thorough analysis of the examination orders of all thirty-seven state medical education programmes that fit the sample since 1990 (data type I) has been conducted. Moreover, an analysis of the discourse on educational accounts since the inception of the JME in 1984 (data type II) offers an additional triangulation point and facilitates a comparison of the incremental developments in the legal and public arenas, following Abbott and Historical Institutionalism (Abbott, 1988; Streeck & Thelen, 2005).

Finally, moving from this meso–level of inquiry to the micro–level allows for more in-depth case descriptions: The field study of the eight Skills Labs in NRW (data type III) illuminates the third arena of the workplace, once again creating a triangulation point from which to see the individual interactions behind the institutional construction of medical education. The validation of fieldwork can again be measured by Maxwell's checklist: Looking for long term engagement (I), rich data (II) and discrepant cases (V), in my case, meant going to the Skills Labs at different times and in different locations for comparability. Respondent validation (III) and reflection on the research interventions (IV) is part of the interview guide's design, as well as the content analysis. This study is not of mixed methods (VI), but involved mixed-data as argued, but there is a numerical side to the data presented (VII), which in turn helps to compare and match the cases by their characteristics (VIII) of old versus new, peer-to-peer teaching versus expert teaching, different types of medical specialty represented or not and so on (Maxwell, 2013, 126–129). All of this helps to ensure that there is a richness of description that stands up to possible comparisons with all of the other twenty-nine medical programmes out of the thirty-seven that have been analysed in this way. As Maxwell puts it himself: "As argued previously, you should think about what particular sources of error or bias might exist, and look for specific ways to deal with this, rather than relying on your selection of methods to do this for you. In the final analysis, validity threats are made implausible by evidence, not methods" (Maxwell, 2013, p. 128). This involves knowing the context of one's data and the limits of its explanatory power, while at the same time getting as much out of it as possible. In thinking about extending the robustness of qualitative studies in particular, it has been argued that historical materials can be greatly valuable because it is possible to do numerical processing in qualitative studies, even with a smaller sample. However, the methods must follow the research question and the practicality of the procedures, aiming for depth of reasoning.

4.3 Summary

In this chapter, the methodology employed in the study was outlined to address the research question concerning SBT's rise as a medical education methodology and its professionalisation. The first half of the chapter reflected on the quality criteria and research context offered by different methodologies and explained why this study can be considered as involving a mixed-data approach. To this end, it draws comparisons with mixed methods methodology, as well as with complex reasoning and triangulation. These approaches offer a pragmatic stance towards the epistemological questions of methodologies, while simultaneously translating quality criteria between different methodological approaches. As a qualitative study, this research provides a meso-level analysis of institutional structures and a micro-level examination of specific case studies. Triangulation also enhances the output. The chapter's second half outlined the study's concrete procedures, including a description of how data from all three types were gathered and processed. The first data type comprises legal documentation that represents the legal arena; the second comprises journal articles representing the public arena and the third comprises participatory observation and in-depth expert interviews representing the workplace arena. Data analysis consists of directed content analysis in all cases and, in some cases, preparatory summative content analysis. The following chapter will present the results of the three professional arenas analysed individually.

5. An Analysis of SBT's Three Arenas

This chapter contains an analysis of each type of data material in isolation. The legal arena is the SBT's most formalised representation in medical education. Accordingly, the differences between the general examination order and individual university curricula are significant, particularly in terms of historical transformation. Additionally, a detailed analysis of the North Rhine-Westphalian (NRW) cases shows that SBT, Skills Labs and other indicators are mentioned early and frequently, thereby suggesting that they have been professionalised to the extent that they are ratified in study programmes. The public arena outlines the agendas of individual stakeholders within the field of medical education by referring to journal articles from the largest German-speaking association. Its transformation outlines educators' organisation towards educational goals, thereby enabling their overall professionalisation. Lastly, the workplace arena reveals the organisation of the individual Skills Labs in NRW, and the distinct separation between teaching and testing practices within SBT. Each arena's singularity only reveals part of the subject's interesting overall outlook; to that end, this work gives the reader an overview of how to better understand the analysis outlined and the singular outcome described. The triangulation (see Chapter 6) will reveal how these three angles of SBT and the introduction of Skills Labs are linked together and how they re-inform each other.

5.1 Legal Arena: Overview of Formal Composition

The initial material to consider for the legal arena are legal papers, which signify medical education's formal development as a profession. In accordance with Abbotts's theory (1988), professions can be regarded as stakeholders within the broader context of society, possessing an interest in shaping their working conditions. One forum through which these jurisdictional claims can be articulated is the legal arena (Abbott, 1988, p. 59). It is, therefore, necessary to consider what the formal control of work in the medical educational system consists of. In Germany, a medical degree is obtained at a public or private university through state examination. The state examination is defined in legal terms in the examination order, which is issued by the Federal Ministry of Health and is therefore binding for all medical educational programmes (see Chapter 3.2). The changes in this regulation and its mention of SBT are a natural first point of entry.

Each university also formulates its own curriculum and study regulations, implementing the examination order, while also defining local areas of focus. Consequently, the thirty-seven state-bound medical faculties were considered as a second point of entry through which to both investigate and compare if and how SBT and Skills Labs are alluded to in their regulations. The two types of legal papers analysed include the amendments to the examination order, issued by the Federal Ministry of Health and in order to understand the nationally binding agenda-setting of medical education in Germany, and the study regulations from each medical faculty that implement the guidelines of the examination order (see also Chapter 4.2, Materials: Legal). Lastly, their different study programmes and particularities, with regard to competence-oriented training and testing, are introduced in order to prepare for the deep analysis of NRW university programmes outlined in Chapter five and six.

Federal Level: Examination Order

The literature on medical education's development in Germany commonly refers to the recently introduced examination order of 2002 that points towards a higher degree of practical training and that mentions competence, specifically social or communicative competence, as an educational goal (Müller, 2012, p. 19). The content analysis's objective was to substantiate this claim by means of a comparison with the development of the examination order and the minor changes implemented between 1990 and 2022, and one additional development from 1979 as a point of comparison. Therefore, mentions of SBT, Skills Labs or related references were actively sought out. Furthermore, the curricula of the thirty-seven state universities in Germany between 1990 and 2022 were analysed, given that the examination order constitutes merely a fraction of the legislative framework that governs medical educational programmes and inasmuch as this information could actually be ascertained (see Chapter 4.2 on the compilation of research materials). A comparison of these two types of legal papers evidence remarkable differences, even before triangulating to the other jurisdictional arenas.

The examination orders are discussed in this first section. Twenty documents were found and analysed within our timeframe using a summative and directed content analysis (see Chapter 4.2). It was found that there is a high level of continuity in the mentions of teaching and testing methods. This includes bedside teaching, classical seminars and lectures and exams with multiple choice options or written parts, all of which were in place during the period considered. However, there is more variation in the content that followed a new version of an examination order where the emphasis was placed on different subjects or specialties.

Nonetheless, only one formal mention of this kind could be found in 2021, with regards to SB. In this instance, an exception to the procedure of practical exams is proclaimed, allowing for the use of simulated patients due to the exceptional circumstances of the COVID-19 pandemic and the avoidance of contact with real patients. In terms of the HI, it can be postulated that the examination orders demonstrate a tendency to lock-in, opting for transformation through Layering, given their requirement to be sufficiently general so as to validate all educational programmes and to ensure legal protection for students and examiners and, thus, change entire formulations or institutional guard rails only very reluctantly, instead adding new details and definitions (Streeck & Thelen, 2005, pp. 23–24). Only an exogenous shock, like COVID-19, could result in a non-permanent activation towards SBT (Streeck & Thelen, 2005, pp. 20–21), displacing traditional examination formats due to the state of crisis. While this activation did not last beyond pandemic times, the overall shift towards competence-orientation in medical education does become visible through the increased importance of practical teaching and testing, for which SBT can be used by the individual medical faculties. The milestones of this shift for the given years of 1979 to 2021 are detailed as follows.

The first examination order considered, from West-Germany in 1979, already has a very similar basic organisation of studies in comparison to contemporary ones, interestingly. The basic studies take six years and an initial first aid training must be completed before starting medical training. The curriculum includes different clinical traineeships and practical training. These practical training sessions take place in small groups; however, unlike the Famulatur today, they do not bridge several disciplines or topics. However, local examinations by universities and the Länder were already established, as were many basic subjects of study. Subjects such as physics, physiology, chemistry, biology, medical psychology and sociology were all covered, although the number of subdisciplines and fields has increased considerably since then, thereby indicating a trend towards specialisation.

The next significant legislative change to the examination order came in 1990, which remained stable in many fundamental aspects of the study, especially when compared to the legislation before (1979) and after (2002). Most importantly, the 1990 document recognises degrees with the same period of study as medical degrees, which is aimed at students from former East Germany and was due to the reunification of the German states in 1989. Furthermore, this document recognises scientific content's substantial growth through the years and redirects the subject towards the item catalogue for state examinations. To this day, the tendency for examinations to become more detail-oriented, due to the ban on recurring questions and the increase of academic work, remains a problem.

The 2002 legislation has remained largely unchanged to this day. All subsequent changes are merely supplements to this examination order, albeit extensive ones. In 2002, longer and more detailed descriptions of the course aims were introduced for the very first time. The practical components and interdisciplinary practice were emphasised, as was the opportunity for students to explore a variety of subject areas. This emphasis was later used by some faculties to argue in favour of SBT's inclusion. However, SBT is not mentioned by name; instead, teaching is referenced in relation to practical training and the number of students observing one patient in bedside teaching is reduced from eight to six. In addition to the internships described, bedside teaching, lectures and seminars and subject-related study groups all became an option for teaching, similar to problem-based learning courses.

The amendment to the 2002 examination order, which occurred in 2012, included the statement that skills and abilities, including medical interviewing, should be a central concern of studies. This was often interpreted as concerning communicative competence, which in turn is a central objective for SBT.

The most recent amendment, albeit only a temporary one, was issued in 2021 as a result of the COVID-19 crisis and not only does it introduce digital teaching formats (p. 1), but it also allows for the use of SPs in examinations: "In justified individual cases, the examination may be conducted with a patient presentation using trained simulation patients" (p. 7, translation J.D.). This changed the state examination procedure for cohorts in 2021 and 2022 in locations that opted not to work with real patients, instead shifting the focus towards SPs and emphasising SBT as an educational concept.

In conclusion, education's fundamental structure remains largely consistent with certain persisting teaching methods, such as bedside teaching and the conventional university courses comprising seminars and lectures. However, SBT's introduction into the examination order in 2021 was driven by the constraints imposed by the COVID-19 pandemic and the subsequent limitation of contact with actual patients. This chapter's second section discusses how individual universities reacted to these legislative changes and the extent to which they incorporated SBT or Skills Labs into their curricula between 1990 and 2022.

Länder Level: University Curricula

It is important to note that while the examination order must be general enough for all medical faculties, individual medical faculties are still bound to their respective universities, which in turn are bound by Länder-level legislation. Consequently, individual medical faculties must give themselves study regulations and curricula that are in line with their respective universities' objectives. As demonstrated, the examination order does not outline SBT, except in a crisis situation like COVID-19; however, individual curricula might still mention SBT. At best, they could address the institutional integration of the Skills Labs into the medical faculties and its subparts, how teaching and testing are organised in reference to SBT, especially if SBT's usage and time spent in Skills Labs is optional or compulsory for medical students. In the event of optional training, insights into student's exact usage would not be possible, but it would indicate that SBT training is welcomed at a given faculty. Furthermore, it would allow for subjects or specialties that introduce SBT to be identified, thereby allowing for a description of when and where SBT has gained ground.

Indeed, a comparison of the study regulations reveals that SBT and Skills Labs are mentioned much earlier than in the 2021 examination order, namely in 2003 in Cologne. However, a discrepancy is also evident in the utilisation of these facilities in teaching, which is not mandatory in many cases. Students can freely choose to spend more time in Skills Labs, compared to exams, and some universities mention the use of simulators or simulated patients. The most common SBT exercises include those for "basic skills" (e.g., hygiene or puncture exercises) and these are mentioned 9 times and for internal medicine (i.e., "physical examination" (8 times)). The most frequently mentioned commodities are SPs (25 times), which are especially prevalent in connection with the practical examination concept of OSCE (Objective Structured Clinical Examination), followed by simulators (10 times), and more precisely as either a simple model/dummies and full-size simulators (high-fidelity manikins) (9 times).

We might consider OSCEs as simulations employed in the context of examinations at the faculty level of a given study programmes. However, this does not directly entail that the Skills Lab was used for this purpose or that mandatory SBT took place beforehand, only that both instances are likely. Nevertheless, twelve out of the thirty-seven locations in Germany explicitly mention a Skills Lab within their study regulations (Regensburg, Ulm, Tübingen, Münster, Marburg, Cologne, Hanover, Halle, Aachen, Bielefeld, Düsseldorf and Dresden). Nineteen out of thirty-seven locations also explicitly mention SBT or its subparts, such as OSCE, within their study regulations (Ulm, Münster, Munich, Cologne, Marburg, Göttingen, Berlin, Düsseldorf, Bielefeld, Augsburg, Hannover, Frankfurt, Greifswald, Würzburg, Bochum, Hamburg, Aachen, Düsseldorf and Dresden).

Returning to the initial query: Is the emergence of Skills Labs at German universities best characterised as Layering, signifying an enhancement of existing medical training techniques, or as Displacement, indicating a substitution of prior medical training methodologies, or perhaps a combination of both? A distinction must be made here between teaching and examination by triangulating the two types of legal papers available; namely, both the examination orders, which are more general, and the universities curricula. The following argument is advanced: Firstly, the introduction of Skills Labs can be considered as Layering because they represent an additional institution in terms of teaching methods. Students and teachers can opt for a more varied selection of teaching formats, as the study regulations demonstrate, but these are not obligatory in all cases. Turning to SBT's formal introduction, it can be argued that there has been a Displacement in terms of examination formats. The examination order is not clear enough about this, because tracing the change in 2021, due to the COVID-19 pandemic, shows that it was not the result of a conscious educational decision that sought to turn against reform inertia, but was instead the result of an exogenous crisis. However, the study regulations do establish a clear connection between exams that evaluate practical skills and the utilisation of simulations. These novel exam formats differ from the oral or written exams that are still being practiced, but with a reduced duration to accommodate the practical exams, such as the OSCE.

The final description of the changes that have taken place can only be achieved through triangulation with two other professional areas. However, the basic transformative arguments can already be formulated such that a differentiation between the levels of education's formalisation, namely examination order on the one hand and the individual curricula on the other, is useful. The following section will outline important transformations that occurred within the eight medical faculties in North Rhine-Westphalia with regard to study programmes and to SBT-related innovations, such as OSCEs.

Case Studies of NRW

In Germany, medical study programmes at universities can be structured in two different ways. While all programmes must conclude with binding state examinations, in order to grant students a medical degree, programmes can either follow the examination order directly, in which case they are called regular programmes, or they can restructure their programmes individually, in which case they are called model programmes, Model study programmes, introduced in the late 1990s, give faculties the opportunity to expand their educational methodologies beyond the minimum requirements of regular programmes. Today, numerous programmes follow a model study programme, offering greater freedom in both the content and format of their curricula alongside the basic state examinations and regulations (see also Chapter 3.2). This is often considered to be a more innovative approach to medical education. In North Rhine-Westphalia, three out of eight programmes are regular and five are model programmes. An introductory comparison will help to identify the reform-mindedness of the five model study programmes in Cologne, Düsseldorf, Bielefeld, Aachen and Bochum. Therefore, initial mentions of model study programmes were sought out in the curricula and indicated that, from that point forward at least, the first cohort of students under that curriculum would gain their state examination through model study course education.

First, Cologne offers options for students to convert from regular study course to model study course in its appendix, which indirectly suggests that Cologne had already implemented a model study course by 2003 (Köln 2003, p. 32). However, the actual wording "model study course" is only headlined in 2021 (Köln 2021, p. 1).

Second, the wording of 'model programme' can be found for Düsseldorf in 2013, although here too it can be assumed that a transition phase began previously, but this is not apparent from the data provided (Düsseldorf 2013_2, p.1). In contrast, the determination for Bielefeld, which was initially implemented in 2021 as a model study course (Bielefeld 2021, p. 1) and for Aachen, which has consistently used the self-description of a model study course since 2003 (Aachen 2003, p. 1), and is more straightforward. However, it should be noted that, in the case of Aachen, the transition phase for students who started before that, and under a regular study programme, would not finish until about ten years later, thereby eliminating any juxtaposition of study programmes. Finally, this phenomenon of transition could be observed in the data for Bochum, where the model programme started in 2005, but both regular and model programmes are described in 2013 (Bochum 2005, p. 1; Bochum 2013, p. 2). All of the other programmes at the eight medical faculties in the North Rhine-Westphalia case study, namely Bonn, Münster and Essen, offer regular programmes that are traditionally closer to the examination order in terms of content.

The implementation of a study model course can be understood as an indicator for the professionalisation of medical education by the given faculties, outlining some of the locations mentioned as early adopters in Germany. However, faculties with regular programmes nevertheless needed to be individually considered so as to outline if and where they might possibly reform their curricula within the realm of the examination order's leeway. The argumentation provided in the public arena will underline that the study programmes are a vital point of discussion for practice-oriented education.

A second indicator, which shall be introduced in the section that follow and will be examined further for its impact in later triangulation, concerns the use of a specific examination format, the OSCEs. OSCEs are an acronym for Objective Structured Clinical Examinations and offer the assessment of practical skills through a parkour-like system of several stations with tasks that students must complete OSCEs (Harden, 1988; Harden et al., 1975). Skills Labs are often involved in their organisation because both simulators and SPs are used for the individual OSCE stations. Furthermore, Skills Labs can also be the physical locations in which OSCEs take place, due to their space and room organisation (Jones et al., 2015, p. 59).

The following list provides information about all eight locations in North Rhine-Westphalia regarding the teaching and testing formats, noting whether SBTs, Skills Labs or OSCEs were mentioned between 1990 and 2022:

- In Aachen, mandatory anaesthesiology training has been taking place since 2004 in the form of lectures, seminars, clinical training in real operating theatres and simulation training (Aachen 2004, p.6, 15). In 2013, teaching formats included lectures, practical training at the Skills Lab, seminars and problem-based seminars. These were complemented by plenary sessions that involved debriefing for practical training and examination courses involving bedside teaching, as well as different thematic blocks (Aachen, 2013, p. 14). In 2018, the Skills Lab and simulated patients were emphasised once again (Aachen, 2018, p. 15). More precisely, practical training in Aachen is described as "knowledge and skills acquired or deepened through activities in the laboratory, the Skill Lab, using a microscope, a cadaver, student test subjects or simulation patients, a computer, and in part through discussion" (Aachen 2008, p. 12; Aachen 2018, p. 15; translation by J.D.). The practical examination comprises the bedside teaching assessment (OSPE) and the simulation-based assessment OSCE as part of the model study course and final examination, following the third section of the curriculum (Aachen 2003, p. 4; Aachen 2004, p. 4; Aachen 2005, p. 4) and which is followed by a subsequent brief definition of OSCEs (Aachen 2003, p. 14; Aachen 2004, p. 15; Aachen 2005, p. 16). This section on the definition and use of OSCEs for determining clinical suitability at the end of the third part of the curriculum was expanded upon in 2007 (Aachen 2007, p. 17), but was not included thereafter. OSCEs are then only mentioned as mandatory assessment tools alongside others, such as written exams, multiple-choice tests, oral exams, laboratory protocols and oral presentations (Aachen 2018, p. 17; Aachen 2013, p. 16; Aachen 2012, p. 15; Aachen 2008, p. 15). This suggests that OSCEs are now well-known enough so as not to require detailed definitions.

- Bielefeld, founded in 2021, has included simulation as part of the oral-practical exams since its inception: "An oral-practical examination lasts between 10 and 15 minutes or between 15 and 20 minutes, depending on the module.' It consists of a simulation, demonstration and/or presentation, followed by an examination discussion. During this, students should demonstrate their ability to apply and reflect on their theoretical knowledge in a given situation" (Bielefeld, 2021, p. 15; translation by J. D.). SBT is once again a mandatory part of the education and assessment process. Teaching formats in Bielefeld include lectures, seminars, laboratory training, workshops, problem-based learning groups, Simulation-Based Training, clinical training involving bedside teaching and ward rounds, as well as self-study (Bielefeld, 2021, p. 14). SBT is specified as taking place within the Skills Lab with simulated patients, peer-to-peer training and simulators to acquire communicative and clinical practical skills (ibid).
- In Bochum, the following assessment formats are mentioned: essays, OSCEs, OSLERs (a bedside examinations), the triple jump exercise (case vignettes/paper cases that students must assess), multiple-choice questions, progress tests (another format of multiple-choice that remains consistent across semesters and student cohorts) and portfolios (a mixture of the above). OSCEs specifically mention the use of simulated patients (Bochum 2005, p. 8). OSCEs can either be used for formative assessment, in which students per-

form specific skills, or for summative assessment, in which students receive grades based on their performance (Bochum 2005, p. 4).

- In Bonn, the assessment of clinical and practical skills is possible by performing on real or simulated patients or by using simulators, commonly taking the form of an OSCE or an OSLER (Bonn, 2018, p. 15; Bonn, 2021, p. 18). In 2022, OSCEs received their own section, including simulated patients and simulators as an oral-practical assessment format (Bonn, 2022, pp. 8–9).
- Düsseldorf mentions its SBT training centre, where practical teaching is supposed to take place in courses for practical skills, such as physical examination techniques, puncture and catheterisation techniques and sonography. Students are also permitted to practise at the centre voluntarily (Düsseldorf 2019, p. 13). SBT with simulated patients for physical examinations and communication skills is mentioned within the section on bedside teaching as preparation for the clinical experience (ibid). Assessment can include different types of written tests, as well as OSCEs (Düsseldorf 2016, p. 8; Düsseldorf 2019, p. 25).
- Essen has mentioned the use of OSCE as an assessment tool numerous times over the years, emphasising its importance and consistent use as an assessment rooted in SBT (Essen, 2022, p. 8; Essen, 2019_2, p. 8; Essen, 2015, pp. 2, 7; Essen, 2014, p. 13; Essen, 2014_2, p. 13; Essen, 2013_2, p. 1; Essen, 2013_3, p. 13).
- The Skills Lab is mentioned as a vital component of longitudinal practical skills education within the description of the model study course in Cologne (Köln 2021, p. 13). Cologne's curriculum's appendix provides a detailed description of the Skills Lab's applications, including training with simulated patients and simulators, OSCEs as assessment formats and its integration into the compulsory curriculum from the first to the tenth semesters for various courses (Köln 2021, pp. 63–65). The teaching Skills Lab should be distinguished from the scientific Skills Lab, which prepares students to work scientifically in the medical field and is not connected to SBT (Köln 2021, p. 66). However, practical assessments in the form of OSCEs are mentioned as early as 2003 (Köln 2003, p. 29).

- In Münster, students must undertake a physical suitability test in order to participate in practical training, including lessons with simulated patients (Münster 2014, p. 5). Therefore, while SP or other SBT are not mentioned directly as an educational practice, this ruling indirectly suggests that SBT takes place in Münster. However, the curriculum's appendix states where and how teaching takes place. For example, the following subjects are taught at the Skills Lab: neurology (p. 26), paediatrics (p. 25), hygiene (p. 24) and surgery (p. 24). OSCEs are also mentioned as assessment tools (p. 16).

This list provides the insight that all locations mention SBT, Skills Labs or OSCEs in some capacity, albeit at different points in time, thereby necessitating further triangulation. Moreover, the distinction between regular and model study programmes is not clearly evident in this outline of SBT's implementation. Finally, the types of other learning and assessment tools are frequently analogous.

Summary

The description of the legal arena alone has already provided two central insights. Firstly, individual universities are quicker to formalise new introductions to their medical education programmes than the ministry is to include them in the examination order. This is evidenced by the earlier mentions of both SBT and Skills Labs. This inertia can be attributed to the examination order's broader scope, which makes it binding for all medical programmes. Secondly, the examination order also demonstrates an evolution towards practical training, thereby allowing the medical faculties a certain degree of strategic interpretation. The section on important transformations for the eight medical faculties in North Rhine-Westphalia, with regard to study programmes and SBT-related innovations like objective structured clinical examinations (OSCEs), determined that the locations mentioned are partly early adopters, in comparison to Germany's general development. In addition, the curricula of all institutions incorporate SBT, Skills Labs or OCESs, thereby indicating a professionalisation of medical education. In conclusion, while the legal arena can delineate the formalisation of specific aspects of SBT, it is not a comprehensive source of information regarding the implementation of educational practices prior to this formalisation or of the underlying rationales for their introduction. Further consideration is required on the public discourse surrounding SBT and on the concrete practice within Skills Labs in order to achieve these objectives, as will be outlined in the following sections.

5.2 Public Arena: Issue Making of SBT

The public arena represents the public discourse of educators concerning the developments of medical education and, thus, provides the analysis with the argumentative positions towards practical training and SBT in specific. Abbott defines the public arena's primary function as involving the legitimisation of the profession or professional group, thereby ensuring that the given professional group can carry out its work as it sees fit (see Chapter 2 and 3.1). In this context, the 'work' refers to both the practices and aspired educational concepts that the educators strive towards.

The articles in the Journal for Medical Education, which is affiliated with the GMA, are interpreted as acting as a conduit for this public arena, given that they represent a claim to social authority (Abbott, 1988, p. 60), while also representing the academic sector with its scientific content. This process can be defined as agenda setting. Publication in this journal entails engagement in the discourse surrounding medical education, which manifests itself in scientific studies about developments in Germany's medical education, as well as numerous practice reports. This section on the public arena will, therefore, provide an overview of the different aspects that the articles refer to with regard to the rise of both SBT and Skills Labs. The large number of documents analysed – four-hundred-and-forty-eight were considered important for fine coding (see Chapter 4.2) – makes prioritising certain aspects over others inevitable. Four sections have been included in this subchapter: 1) the JME's introduction, 2) Skills Labs and their commodities, 3) SBT's formal organisation 4) and the didactical concepts as public issue.

The Introduction of the Journal for Medical Education

The following section provides the history of the JME as a whole. This helps to put the JME in context and outlines its significance for medical education in Germany and beyond. The JME's current publication format is Open Access via the research platform German Medical Science (GMS), which hosts several scientific journals and is funded by the German Research Foundation (DFG) and the German Federal Ministry of Health and its sub-institutions (Federal Ministry of Health & German Research Foundation, N.N.). The full name of the journal in question is the "GMS Journal for Medical Education", but the initial name was only the German version "GMS Zeitschrift für medizinische Ausbildung". This notwithstanding, the journal and its content are published by the GMA and are independent of the Ministry or other institutions. The GMA was also founded as the German branch of the AMEE, the Association for Medical Education in Europe (1984-H1 (4)). This affiliation is emphasised in the JME's inaugural issue in 1984 through the incorporation of a letter from the then-director of the AMEE, Henry Walton. In this letter, Walton congratulates the establishment of the JME as a "great event" and acknowledges its contribution to medical education in European countries (1984-H1 (9)). The GMA, and JME as its publication organ, have certainly become an influential stakeholder for medical education throughout the German-speaking world, as evidenced by the union of the MFT and the NKLM (2016-V33-I1 (7), P. 3).

Individuals in leadership roles drive project progress; To that end, the JME's inception will be outlined in the triangulation of the public and workplace arena. Dietrich Habeck, a professor in Münster involved in SBT implementation, was also the JME's chief editor in its early years (see Chapter 6). The timing of the journal's implementation, six years after the GMA's introduction, is also significant. The journal's growth is intertwined with the GMAs, as evidenced by GMA committees addressing diverse subjects in medical education and regional groups representing local networks (2012-V29-I5 (8), P. 1). However, the JME cannot be comprehended as a journal for German-speaking medical education exclusively, despite this being its primary function. Its international orientation is evident from its very inception, as evidenced by its establishment and affiliation with the AMEE, as described previously. This early commitment to a global perspective is further emphasised by the occasional presence of publications in English (1987-H1 (1), P. 2). There are further instances in which SBT is quoted as being further advanced on the international scale in comparison to Germany. With regard to Skills Labs, the Netherlands is quoted on multiple occasions as a role model, with Utrecht and Maastricht in specific being cited (1986-H1 (2), P. 4). This prompts the question of how the development of Skills Labs in Germany is represented within the JME articles.

Figure 5: Congratulation to the founding of the JME (1984-H1 (9))

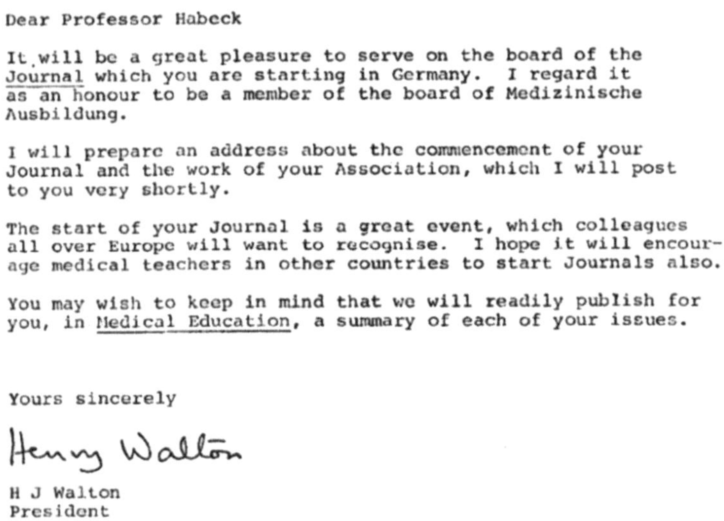

Dear Professor Habeck

It will be a great pleasure to serve on the board of the Journal which you are starting in Germany. I regard it as an honour to be a member of the board of Medizinische Ausbildung.

I will prepare an address about the commencement of your Journal and the work of your Association, which I will post to you very shortly.

The start of your Journal is a great event, which colleagues all over Europe will want to recognise. I hope it will encourage medical teachers in other countries to start Journals also.

You may wish to keep in mind that we will readily publish for you, in Medical Education, a summary of each of your issues.

Yours sincerely

Henry Walton

H J Walton
President

Skills Labs and Their Commodities

As demonstrated in Appendix 5, sixteen out of thirty-seven medical faculties explicitly report on having a Skills Lab or similar facility for SBT, including descriptions of their founding procedure or working practices. Instead of discussing singular Skills Labs and its articles, they are instead summarised by a few sentences in Appendix 5. This finding also indicated a notable interest in SBT by the JME's readers, as well as an opportunity to analyse SBT's advantages and disadvantages according to the experts.

An increased number of meta-studies on Skills Labs can be found throughout the years. For example, an article from 2018 explores the rationale behind setting up a Skills Lab from a risk management perspective. Surprisingly, interviewees in this study were often unable to provide a concise or uniform answer as to why Skills Labs or simulation centres are useful (2018-V35-I3 (11), p. 1). This demonstrates a lack of standardisation, which is often criticised (2018-V35-I3 (11), p. 3). Another article, from 2016, uses a literature review to examine the advantages of Skills Labs and their implementation since the 1980s. Like my own analysis, it hints at the standardisation of the learning environment and learning goals, the possibility of providing feedback and the issue of fidelity. This leaves SBT as a starting point for practice, but not a substitute for clinical experience (2016-V33-I4 (11), pp. 7–8). Moreover, the modus operandi of Skills Labs is oriented around its commodities, as outlined in Chapter 3. Consequently, the three categories of commodities – namely simulators, SPs and commodities on the verge – are also a recurrent focus of the JME articles about SBT.

Simulators

The following articles represent the variety of ways in which simulators can be defined and found, considering their connection to specific subdisciplines of medicine as *intra*professional education and as representation of the clinical environment as *inter*professional education. The organisation of simulators into three main categories (see Chapter 3.3) originates in the examples given within the JME articles, alongside the description within expert interviews and SBT (see Chapter 5.3). What is noteworthy is that the degree of ability may align with the historical development considering their production (i.e., materials and technology needed, but not with the transformation of use). All three sub-formats of simulators, namely simple models, simple dummies and high-fidelity manikins are all in use today.

For example, can a contemporary 3D-printed heart be considered a simple model? It embodies the anatomy of hearts of varying sizes and represents adults, teenagers and infants in order to assist with comprehending the developmental stages of a heart's growth. Simple models are designed to represent specific anatomical features in a limited capacity. Another widespread example of a simple model used in SBT is suture pads used to learn stitches, a basic skill learnt within the first semester of medical studies (2010-V27-I1 (1), p. 3).

An intubation torso is understood as an example of a simple dummy. This simulator possesses a range of functions that extends beyond those of the basic model because it is capable of demonstrating the flow of air and liquid while ensuring a person's (dummy's) airway (JME Article 2010-V27-I1 (1), p. 6). However, this embodiment is also limited in its functionality, given that it only displays a head and torso. Similar simple models have been documented at German faculties since the 1980s, with the "Thoraxtrainer" from Münster serving as a notable example. This dummy is utilised for auscultation training, producing sounds that simulate the symptoms associated with different heart failures (1986-H1 (2), P. 2; 1994-H1 (1), P. 1).

Finally, and in direct comparison to the limited Thoraxtrainer, stands the high-fidelity manikin. An example from Leipzig examined the effectiveness of training in their SBT and Skills Lab with a "life-size cardiology patient simulator (that) has been introduced and its value for teaching auscultation was tested with good results" (2021-V38-I6 (8), p. 1). In this example, auscultation and the output of heart sounds are once again central, but this manikin is life-size and more technologically advanced than a simple dummy because the overall abilities reach a higher level, which may also make these manikins more vulnerable to technical malfunctions, but showcases technological components that extend beyond the manikin, such as screens or monitors, thereby providing supplementary data.

It is noteworthy that the GMA's Practical Skills Committee created an internal database and rating overview about different simulators and their use in Skills Labs in German-speaking countries. This private homepage functions as an encyclopaedia with individual pages dedicated to the given simulators, concise descriptions of the individual Skills Labs and the description of use in teaching, according to self-description (2013-V30-I1 (4), P. 2–3).

SPs

SPs do receive a high number of mentions within the journal articles examined. They are most often cited together with the increase of curricula parts concerned with communicative or social competence, even though they can also be used for hands-on training of specific practical skills. In 2015, almost all medical faculties stated to work with SPs (2015-V32-I5 (7), P. 6). This can be understood as an expression of the expectation that a medical professional will be able to both comprehend and articulate their academic education's abstract knowledge into everyday language. Moreover, it is not solely the abstract medical knowledge itself that renders communication between doctor and patient a sensitive matter, but also the diagnosis that the doctor is obliged to communicate. During their studies, students are introduced to a variety of communication models and theoretical tools, including those that address the difficult task of breaking bad news (2011-V28-I4 (4), p. 2). However, the faculties and, within them, the Skills Labs take their cue from the medical education system in order to implement these changes within the curriculum and to focus on learning objectives. As stated in Chapter 5.1, the formal regulation of the examination order does not offer a framework for the introduction of (longitudinal) communicative skills within the curriculum and it also does not provide guidance on the use of SBT with SPs. Consequently, the NKLM was developed as a form of quasi-regulation and is employed by educators in this public arena as a legitimation for this teaching format (2021-V38-I4 (9), P. 3).

In any case, it is evident that the efficacy of students' competencies in the SPs training programme must be assessed as either successful or as requiring enhancement. This opens up two aspects: Firstly, with what measures does this assessment occur, and secondly, who is responsible for providing this assessment? Two kinds of tools can be used as measures as additional orientation for debriefing: video footage and observation scripts. Video footage can be defined as the synchronous transmission from the simulation room to the debriefing room, thereby allowing for greater spatial separation between the two rooms, depending on the Skills Lab's structure. Alternatively, video footage can be stored digitally and rewatched by students to reflect on and learn from past SBTs. An exemplary report of this practice can be found in Mainz (2021-V38-I3 (16), P. 8), which details the use of video recordings of SBT with SPs. The second measure has been shown to help students concentrate during observations. As outlined in the SBT process, a limited number of students usually participate directly in the simulations, while more observe and subsequently debrief. In this pilot study, conducted in Tübingen, students who used a collaboration script reported improved observational skills, along with self-reflection (2018-V35-I3 (7), P. 1). Students, SPs and educators often give feedback in teaching. However, in a testing scenario, feedback is provided by educators, clinical lecturers or by SPs (2021-V38-I4 (8), P. 4).

It is noteworthy that the JME has described the use of simulated patients since its implementation in 1984, with the observation that the increase in student numbers makes the presentation of patient cases in the clinical environment exclusively insufficiently viable to providing all students with an opportunity to train. Two locations, Bonn and Münster, are cited alongside Dutch Skills Labs as practical examples with a positive experience (1984-H1 (3), P. 1, translation J.D.).

This early article, written by Dietrich Habeck, outlines advantages and disadvantages of SPs and is still relevant today. These include the mitigation of stress on actual patients, the capacity to train extraordinary cases, the standardisation of training environments, the evaluation of students in a non-intrusive manner and the availability of SPs on demand. However, there are also several disadvantages, including the artificial nature of the training situations; yet, the effect of the critique of which is reduced by the distinction between the artificial simulated setting and the less artificial communication that nevertheless takes place between student and SP. The limited capacity of SPs to represent a wide range of symptoms is also mentioned, but is thought to be similar to the constraints encountered in clinical settings more generally (because not all symptoms are represented) (1984-H1 (3), P. 3–6).

SPs have become a widely utilised component within the framework of SBT in the period spanning from the 1980s to the 2020s. A comparative analysis of SP programmes in German-speaking countries reveals that significant variations exist in organisational and didactic methodologies, even though similarities can be identified, such as SPs providing feedback during the debriefing phase of SBT, (2019-V36-I3 (4), P. 1). Consequently, this section advocates for enhanced standardisation in SP programmes to ensure the quality of education and to strengthen their role as effective tools for practical examinations (2019-V36-I3 (4), P. 1). In this regard, the subsection of the GMA on simulated patients includes a statement on the minimum standards of SP programmes in the JME same issue (2019-V36-I3 (5), P. 1), which will be considered further below under regulations and within the triangulations with the legal arena in Chapter 6. In conclusion, the articles under consideration in this section on SPs regard their role as a central aspect to SBT and as a valuable asset to Skills Labs at German faculties. This is achieved through their organisation in Simulated Patients Programmes and their involvement in the assessment of students. At the same time, the articles allude to the latent competences, especially in communicative skills that medical professionals need to possess.

Verge

Different materials can be summarised as being on the verge of SBT. Firstly, it includes instruments and consumables that are part of the training, but not necessarily specific thereto. Secondly, it includes simulators that are self-made, which is not common, but does showcase specific historical circumstances in which commodities were not purchased, but were instead made by educators. Thirdly, it includes technology that might be part of SBT in the future, but which is not sufficiently developed to be used widely, VR technology for example.

Consumables include clinical materials that are handled routinely in SBT. Blood collection classes include the cannula and how to both open and handle it safely. Wound care requires the correct use of suture needle, suture thread and bandages. One Skills Lab reused materials for SBT that had expired, thereby reducing costs. Specific medical instruments are central to practical skills and this not limited to SBT; one example is operating and reading an electrocardiogram. All three illustrations can be found in an article about Dresden's Skills Lab (2010-V27-I1 (1), p. 3).The literature also contains examples of educators creating their own SBT commodities, often while introducing new skills and materials. The report on Dresden's Skills Lab describes the founding phase as economically restricted, which led educators to build and patent their own simulators, including simple models for blood collection (2010-V27-I1 (1), p. 6). Other reasons for adjusting common SBT commodities include the COVID-19 pandemic, which entailed a critical juncture and changed teaching from face-to-face to online and self-learning, due to the necessity to avoid contamination. One report from the Technical University of Munich describes how students learned the practical skill of inserting a urinary catheter by building their own simple model using everyday materials, with positive feedback from students about being able to continue hands-on practice even if they were unable to attend classes at the Skills Lab (2022-V39-I1 (4), p. 1).

Another way of training the movements required for practical skills aligns with the latest introduction of commodities in SBT. This includes VR and AR that similarly facilitate learning experiences by focusing on bodily experience and that foster professional growth. The literature provides examples of VR and AR programme development, but most are not yet used in mainstream education (2019-V36-I4 (1), p.1; 2020-V37-I6 (5), p. 5–6).

Formalities of SBT

The following section is dedicated to the formalities of SBT and Skills Labs as described in the JME articles. Accordingly, the OSCEs as practical examination option are discussed, followed by the medical faculties' composition, in which the Skills Labs are institutionally embedded, and lastly the debate on regulations and quality standards of SBT in medical education in Germany.

OSCEs

OSCEs are often mentioned alongside SBT, due to their organisational challenge and their importance in practical examination, in the JME articles. The following examples provide an overview of the importance of OSCEs and related arguments for and against SBT. OSCEs are often developed around specific curriculum points. For example, stations can be designed to resemble rooms, correlating with students' experiences. In Marburg, this connection is used for general medicine, with one station being dressed like a doctor's room in a private practice. This takes place after students' private practice placement, so they should be familiar with the circumstances (2016-V33-I4 (4), p. 2). This shows how theory and practice can align in curricula. Another report from Heidelberg argues that practising OSCEs does not necessarily improve practical skills, but the prospect of being tested in the OSCE exam increases the hours that students use the Skills Lab for the purposes of training. This shows the possibility of a hidden curriculum: what will be assessed gets trained beforehand (2012-V29-I5 (5), p. 9).

The tool of assessment exerts a significant influence. A 2018 study compared multiple-choice questions and OSCEs as assessment tools, finding that each fosters and measures different skill sets. The OSCE may impact the educational process and can support the development of skills and behaviours required for clinical practice, while the multiple-choice modality fosters the acquisition of knowledge (2018-V35-I4 (7), P. 5). OSCEs are complex and expensive, with personnel costs covering the Skills Lab and clinical examiners. A positive aspect of OSCEs is that examiners reflect on the expectations of students and the medical education overall, resulting in improvements to training concepts. A 2011 pilot study in Ulm provides a concrete cost statement with overall personnel costs of about €12,500, spread over SPs, student assistants, examiners and educational staff and the three phases of planning, implementation and post-processing (2011-V28-I1 (9), P. 12). This does not include indirect costs for rooms and commodities; these can be expected to be higher currently.

The first documented instance of OSCEs being employed at a German state university within the JME articles was in 1995 in Hannover. Initial trials in 1993 and 1994, involving 20 and 80 students respectively, proved successful, albeit complex, with a high level of individual attention being required from the examiners. The response from students was favourable, with OSCEs being regarded as an effective method for assessing practical skills (1995-H1 (15), p. 1). In this context, assessment is defined as formative, providing students with the opportunity to both understand and improve upon their weaknesses, as opposed to summative assessment that only provides a pass/fail result (1995-H1 (15), p. 8). OSCEs became more established after the 2002 examination order, much like with the development of the SBT as a whole. Their growing significance in the assessment of SBT has prompted educators to focus on quality standards, regarding the SPs' and examiners' instruction. This is particularly evident in considerations of the integration of OSCEs and SBT commodities into the state examination's high-stakes practical exams (2020-V37-I4 (2), P. 10). This examination order has the potential to significantly impact medical education, but it has not yet been implemented.

In conclusion, the OSCEs have been demonstrated to be central to SBT. The importance of Skills Lab as an institution that organises these assessments can be argued for persuasively, given that the teaching of practical skills is an integral component of the educational process. However, this raises the question of how educators evaluate the reform inertia of regulations, specifically concerning the examination order.

Regulations

As suggested previously, the main problem for any educational programme, but especially when considering the assessment of practical skills, is how to construct the measurement of performance and what is considered successful performance in the first place. Knowledge and know-how can be considered as different entities in this regard (2005-V22-I4 (13, P. 1), similar to Abbott's three-step analysis of professional behaviour of diagnosis, inference and treatment, with inference being central to know-how because it involves the consideration of different solutions to a given problem (Abbott, 1988, pp. 40–52). Educators naturally consider analogous problems in the context of reforming medical education within the JME articles.

As early as 1986, the report of the GMA meetings called for an experimental clause in the examination order in order to give the faculties the opportunity to experiment with a better integration of the medical curriculum's preclinical and clinical parts (1986-H2 (2), p. 5), but it was not until the 1990s that the reform or model study programme became an option for medical faculties (see Chapter 5.1). In 1995, the GMA itself issued a survey about innovations in medical education that the different faculties strive towards and published a summary about it. This commentary described the importance of medical education as rising, but deficient in terms of knowledge about educational literature or exemplary projects in Germany and abroad (1995-H2(5), p.1). In a similar vein, given that the JME issues functioned as documentation of the GMA conferences, the printing of a speech from 2001 critiques the missed opportunity for a fundamental reform of medical education with the – not yet in place at the time – 2002 examination order. The early discussion phase included the removal of the pre-clinical and clinical phases and the introduction of longitudinal and problem-oriented learning, not only with real patients (which argues in favour of SBT), but which were dismissed in the final version (2001-H1 (12), p. 2). This caused "disappointment" (ibid, translation J.D.) among reform educators. It is also noteworthy that the MFT is directly attacked for its reform-blocking agenda with and its rejection of model study programmes (2001-H1 (12), P. 2). This demonstrates not only that legislation is subject to multiple rounds of revision, but also that the public arena exerts significant influence on this process by either advocating for or rejecting draft legislation, as is publicly documented in this article. The triangulation of the public and legal arenas, as outlined in Chapter 6, will provide a comprehensive account of the agenda-setting and the back-and-forth between the GMA and other associations in the Federal Ministry of Health, as the governmental issuer of the examination order. However, the current structure of medical curricula is characterised by significant heterogeneity and this because the regular and model study programmes are still in place. These programmes can, in accordance with the examination order, use gaps and leeway to implement reformatory parts within their curricula.

Consequently, it can be argued that the implementation of a model study programme provides singular faculties with the opportunity to reconsider their curricula. In the period between 2000 and 2022, a number of other locations decided to implement model study courses, a development which is also reflected in the rising number of articles featured in the JME on the subject. In 2007, only six state universities (Berlin, Aachen, Cologne, Bochum, Hanover and Mannheim) were identified as offering model study courses, with the article emphasising that other locations had also reformed their curricula within regular study courses. Two of the aforementioned locations (Berlin and Bochum) had introduced a two-track system, allowing students to choose between the two study programmes. The establishment of a model study course is a work-intensive and prestigious undertaking for the respective universities (2007-V24-I3 (5), P. 1). In comparison, an overview article from 2019 identifies nine state universities with model study programmes, including Hamburg, Düsseldorf and Augsburg, and provides numerical data for each implementation: Berlin (1999), Aachen (2003), Bochum (2003), Cologne (2003), Hannover (2005), Mannheim (2006), Hamburg (2012), Düsseldorf (2013) and Augsburg (2019) (2019-V36-I5 (15), P. 5). This is a factor that requires consideration in the subsequent analysis of the locations in North Rhine-Westphalia, given that five of the eight locations (Aachen, Bochum, Cologne and Düsseldorf – and Bielefeld not yet founded in 2019) have a model study programme, in contrast to the other three that have regular study programmes (Münster, Essen and

Bonn). Furthermore, SP-programmes and Skills Labs are a central concern, even in the initiative phase of implementation, when describing reformed curricula towards a competence-orientation and problem-based learning (2011-V28-I2 (6), P. 2).

It is not solely singular faculties that are undergoing a transformation in their formal structure towards competence- and a practice-oriented education. Rather, it is the alliances throughout all of Germany that are adopting a reformist stance, resulting in the establishment of the NKLM (see Chapter 6: Triangulation for a detailed analysis of the interplay between the legal and public arenas). This shift is also explicitly articulated within the JME articles, and not only applies to the restructuring of curricula towards students' skills, but also to the educational abilities that educators and clinical teachers offer and that needs further reflection (2017-V34-I4 (6), P. 1).

The study found that significant changes are possible and the willingness of teachers to align their lessons towards competence-orientation is given, but will not be implemented without outside impulses, due to time pressure and workload. Therefore, overall faculty development is central, to avoid a hidden curriculum of known contents (2017-V34-I4 (6), P. 8), which entails an institutional restructuring that can be considered as Displacement (Streeck & Thelen, 2005, pp. 19–31). The triangulation with the workplace arena shows that this change is not as thorough as the advocates argue as will be demonstrated in Chapter 6. However, the public arena is strong and direct with their claims towards reform, possibly because they are aware that it is more efficient to start strong and to then let go of some aims in further negotiations in the legal and workplace arenas. This function of the JME, as a public forum for such claims, can be traced throughout the given timeframe.

The Federal Ministry of Health has been the subject of consistent criticism in the GMA's comments. However, during the 2010s, the discourse shifted from criticism of the proposed drafts for reforms to the educators' own proposals towards competence-orientation and, with it, the expansion of SBT. A central proposal of this kind is the consensus statement made by the GMA Practical Skills Committee in 2011. This paper is the result of an initiative by the directors of Skills Labs in German-speaking countries since 2007 to found the given committee within the GMA. Its aim is the improvement of implementation of practical skills through defining specific competencies and skills which should be required by students throughout their studies (2011-V28-I4 (9), p. 8).

Figure 6: Activities of the GMA Committee Practical Skills (2011-V28-I4 (9), p. 9)

Table 1: Chronology of the consultations for the competence-based learning targets catalogue of the GMA-Committee on Practical Skills.

Consultations for the competence-based learning objectives catalogue of the GMA-Committee on Practical Skills	
Targets Catalogue Start	Decision by the GMA Committee on Practical Skills at the Spring meeting of the Skills Labs (Cologne 2008)
Learning Targets Table	Compilation of existing learning objectives catalogues on practical skills in the German speaking world: Berlin, Düsseldorf, Erlangen, Frankfurt, Freiburg, Göttingen, Kiel, Cologne
1. Review Process (sort)	Determining the new approach in the GMA Committee on Practical Skills at the Spring meeting of the Skills Labs (Aachen 2009) and formation of a task force: Aachen, Cologne, Kiel
2. Review Process (condense)	In collaboration between the Skills Labs in Aachen, Frankfurt, Bern, Cologne, Düsseldorf, Erlangen, Dresden
3. Review: Panel of Experts	Meeting and agreement on AP 14 of the NKLM by MFT / GMA (Frankfurt 2010): Processing and circulatory feedback
4. Review: Focus Group Discussion on the Learning Targets Table	GMA Committee on Practical Skills (GMA Annual Conference 2010)
5. Review	GMA Committee on Practical Skills at the Spring meeting of the Skills Labs (Würzburg 2011)

This resulted in a list of 289 different practical learning goals assigned to 12 organ systems, with three overlapping areas to other fields and one area of across-organ system skills (2011-V28-I4 (9), p. 7). This list is based on learning catalogues from German faculties and the Swiss model, while a similar catalogue was also implemented and made compulsory in Austria in 2011 (2011-V28-I4 (9), p. 7–9). The NKLM, the result of a collaboration between the GMA and the MFT, was initiated, but was not yet finished. It was published in its first version in 2015 after six years of development (2016-V33-I4 (10), p. 2). However, the Consensus Statement already references the NKLM as being insufficiently comprehensive when compared to the list provided. The consensus statement can, therefore, be seen as central and preparatory to the development of competence-based education in Germany, and it puts SBT centre-stage for both assessment (i.e., with OSCEs) and in teaching (i.e., with simulators) (2011-V28-I4 (9), p. 10).

The consensus statement's conclusion almost reads like a manifesto and prophesises changes in German medical education between 2011 and 2022. To return to the argument of critique of the Federal Ministry of Health, the ministry as well as the IMPP as the assessment institution thereof, are called out about the competition-inducing format of the written State Exams, instead of furthering the reform of medical education towards practical competence, even though the 2002 examination order did give leeway to implement practical teaching more than previous regulations had (2011-V28-I4 (9), p. 8). In the following, the consensus statement is used for the evaluation of given curricula, thereby providing further importance to its flagship function and, consequently, to the NKLM (2012-V29-I4 (7), P. 6).

However, the Federal Ministry's reaction is visible even within the JME articles. For instance, an analysis conducted by the IMPP draws comparisons between the then-current exam catalogue (Gegenstandskatalog) and its learning objectives and to the comparably newly introduced NKLM (2017-V34-I1 (7), P. 1). To consider that an institution so central to the assessment of medical education, providing the exam formats for the state examination according to the examination order, compares its exams to the NKLM shows how influential the latter is. At this point in time, the Federal Ministry of Health and its institutions were under pressure to react to the immense effort and self-organisation by the educators represented in the NKLM, given that they developed this quasi-legislation as a union between different educators' associations and did so in addition to their assigned workload (i.e., working groups of educators and medical teachers holding meetings in their free time to discuss wordings, organisation and argumentation towards competence-orientation). The study finds that the exam catalogue and the NKLM show a high congruence of content, although they admit that the two documents' wordings are different and, therefore, are difficult to compare in the first place (2017-V34-I1 (7), P. 4 – 5). Ultimately, the NKLM is evaluated as a valuable resource for faculties to use for their curricula reforms, given that it is backed by these influential institutions, suggesting that the examination order could also be inclusive about the NKLM's aims.

In 2017, an expert commission was established, and the Federal Ministry of Health published the reform plan, entitled "Masterplan Medizin 2020" in 2018. This plan proposed many recommendations aimed at promoting competence-orientation and restructuring curricula. In response, the GMA swiftly formed a commission to evaluate the Masterplan and to consolidate the various GMA committees' positions into a single document. The evaluation, published in 2019, reiterates the initial response to the Masterplan as positive, but unfinished (2019-V36-I4 (11), P. 1). The Masterplan is conceived positively, especially with regard to SBT, but the lack of definition of central concepts (such as competence) is criticised and the integral part of medical faculties in the development of reforms is emphasised. In the GMAs argumentation, many faculties have already implemented many of the Masterplan's demands (2019-V36-I4 (11), P. 4). Furthermore, the report underlines that "this commentary is not a political statement on the content of the Masterplan for Medical Studies 2020 but rather a contribution to the scientific discourse about its contents" (2019-V36-I4 (11), P. 1).

This is remarkable because, according to Abbott´s account, the statements made within the public arena can always be considered as policy-inducing and agenda-setting, due to the jurisdictional claims being made and defended. According with both Abbott's and Thelen's theories significantly, the medical educational professionals and the medical faculties they work in do not wish to reduce their freedom institutionally (see Chapter 3), calling upon their sovereignty, which is historically strong in Germany (Simon, 2021, p. 17). The GMA welcomes the Federal Ministry of Health's commitment to implementing measures nationwide, but demands "suitable funding" for this nation-wide endeavour (2019-V36-I4 (11), P. 4). This approach could be regarded as a bold strategy to resist governmental influence over academic structures, while emphasising the importance of financial support from their sources. However, this approach aligns with the historical models of universities and state institutions in Germany, where a strong emphasis on autonomy and self-governance has been a characteristic feature.

Furthermore, the educational community must be credited for its continuous internal debate on the standards of work, which demonstrates a high level of self-governance. Faculty development is a fundamental objective in the context of medical reform. Consequently, it should be regarded as a crucial component in SBT's implementation. The rationale for considering medical faculties as a management instrument within the institutional context of medical education was clearly articulated in a position paper by the GMA as early as 1993 (1993-H2 (3), P. 3, translation J.D.).

By the 2000s, the idea of achieving a consensus on quality standards in medical education had gained momentum, further accelerating SBT's establishment (2007-V24-I2 (14), p. 6). Standardisation has also become a prominent theme in the literature about SBT and Skills Labs since the 2010s. This can be attributed to SBT and Skills Labs having been well-established by that time. Consequently, there has been a shift towards both targeted evaluation and standardisation. However, the formalities of implementing SBT and Skills Labs represent merely one aspect of the rationale. The usefulness of such training in a didactic sense is equally, if not more, significant.

Didactics as Public Issue

The underlying rationale of the public arena on medical education transformation must naturally consider educational standards and theories, not only towards their formal implementation in curricula and institutions, but also beforehand with questions of orientation of education, the practicability of trainings and the standardisation of assessment.

Once again, this study does not assess the different educational approaches towards their usefulness, but instead traces the changes in stances formulated within the medical educational profession's three jurisdictional arenas. One of the issues voiced in the public arena is the agenda-setting about what kind of education should be aspired to and what the problem with given education is in the first place.

Theories of Learning and Learning Tools

As mentioned previously, the orientation on international role-models from the very start of the JME was prevalent. The learning models that would later be considered ground-breaking in model study courses, namely combining the clinical and non-clinical sections of the studies and introducing *problem-based learning* in specific, are quoted as aspirational for German medical education (1986-H1 (2), p.11).

This case-based learning also lays the fundament for a second learning paradigm, which is *lifelong learning* (1987-H1 (2), p. 10). These demands respond to a growing problem identified by educators in the 1980s, namely the drifting apart of the theoretical and practical parts of educational programmes and this is attributed to the increasing amount of theoretical knowledge that cannot be memorised in its entirety, instead students need to *learn how to learn* and to keep their knowledge up to date after their studies (1988-H1 (5), P. 2). An article from 1986 estimated that specialist medical knowledge can be considered outdated after five to ten years, concluding that a fact learned at the beginning of a student's studies would *be outdated by the end of their education* (1986-H1 (2), pp. 15 – 16), which can be considered to be even more relevant today.

Furthermore, to return to the implementation of more practical teaching, the problem of the number of students in relation to the number of patients has been increasing since the 1990s (1990-H3 (2), P. 1, translation J.D.). How students are and can be *involved in patient care is controversial nowadays*, given the high standards of patient safety and the changes made to the healthcare system towards outpatient care, together with the *ever-increasing number of students*.

Finally, when considering the ways in which students learn practical skills, one concept predominantly cited until today is *Miller's theory of assessment of clinical skills/competence/performance*, originally from 1990. In brief, it contains a rising degree of independence among students from "knows" something (knowledge), to "knows how" to do something (competence), to "shows how" to do something in a training situation (performance), to "does" something on the actual patient (action) (Miller, 1990).This theory is often used to advocate for SBT, but also for other practical skills training (2009-V26-I1 (9), P. 3).

However, the concept of practical skills remains constant, even after having considered different learning constructs since the 1980s and the problems faced by medical education in this very concise section. Although all of the concepts mentioned co-exist today, the question of definition of practical skill in medical education persists nevertheless, (2016-V33-I4 (12), P. 2). This raises the question of how to assess practical skills, including aligning learning goals with learning and teaching modalities, such as location and commodities (2009-V26-I4 (3), p. 15). A problematic aspect of restructuring curricula in accordance with such outcomes is the temporal discrepancy between the planning of education and the anticipated changes in the future working field of medicine, running the risk of producing outdated curricula by the time that they are implemented (2016-V33-I4 (10), P. 2).

This argument about knowledge's short shelf-life was mentioned previously, but will be extended here to include professional skills. Therefore, SBT offers and approaches medical education through a variety of commodities and modes of training, whether it be with simulators, SPs or peer-to-peer teaching, all of which offer advantages and disadvantages that are discussed throughout the literature (2016-V33-I4 (12), P. 1; 2008-V25-I1 (16), p. 1; 2002-H2 (35), P. 6) The learning goals, the contents to be addressed in SBT can be as elementary as the correct sanitisation of hands, are a matter that falls under the domain of clinical hygiene (2019-V36-I4 (5), P. 8) and are as complex as teaching and testing the ability to work in a team structure, which is a given in future clinical work (2015-V32-I2 (6), P. 10; 2020-V37-I5 (6), P. 10).

To further underline the analysis of the public arena, this short section provides an overview of the codes used for the analysis, before concluding with a summary. It is important to note that, while the examples mentioned in this section may appear arbitrary to the reader, they are the result of a multi-step process (see Chapter 4). It would have been impractical to quote all of the relevant articles; the review already covers an extensive range of topics. Nevertheless, each article represents an argument about SBT, Skills Labs or related topics.

The quantity of absolute document numbers mentioning specific topics provides data on the contents and therefore a hint towards increases or decreases in relevance over time. Firstly, the function of SBT, towards teaching and testing, has been emphasised repeatedly. A code of the same name provides a proportion of the mentions: the function of SBT for education is formulated directly in 53 articles, 35 of which formulate evaluation as examination or debriefing and feedback as the function of SBT, while 30 formulate it for training and learning purposes. This means that several articles mention both functions at the same time, while the overall distribution suggests an almost split decision between the two functions.

Finally, the tables in Appendix 5 present a summary of the advantages and disadvantages associated with SBT. Once again, the absolute numbers of documents provide an indication of the relative importance of the various issues. While each aspect requires separate consideration, particularly concerning patients' safety, which is mentioned twenty times as an advantage, and the effort and costs in preparation of SBT, which is also mentioned twenty times as a disadvantage, and represents the basic pressure SBT that is often under and what it (says it) offers in comparison. Finally, a description of the rise of the term "Skills Labs" was undertaken because these codes do not differentiate between the timeframe in which these stances are taken.

The two graphs of figure 7 show the absolute number of documents in which the concept of Skills Labs is mentioned over time, suggesting a transformation of its importance. This representation includes other terms such as simulation centre or training centre, or the description of rooms in which SBT takes place and can, therefore, be considered as a broader concept than the term Skills Lab itself, even including its various lemmas. The upper graph, representing JME articles, shows an almost constant level of growth, with a rapid doubling of the numbers.

This rapid increase between the years 2008–10 with 10 mentions to the years 2011–13 with 21 mentions could be explained by the increased focus on competence-based education with the Skills Labs initiatives at its centre (i.e., the Consensus Statement issued by the GMA Practical Skills Committee). This growth falls back to its previous level of 10 mentions in 2017–19, possibly due to the shift towards a competency focus within medical education, shifting attention away from Skills Labs. Further research is needed to explore this.

Figure 7: Rise of the term Skills Labs

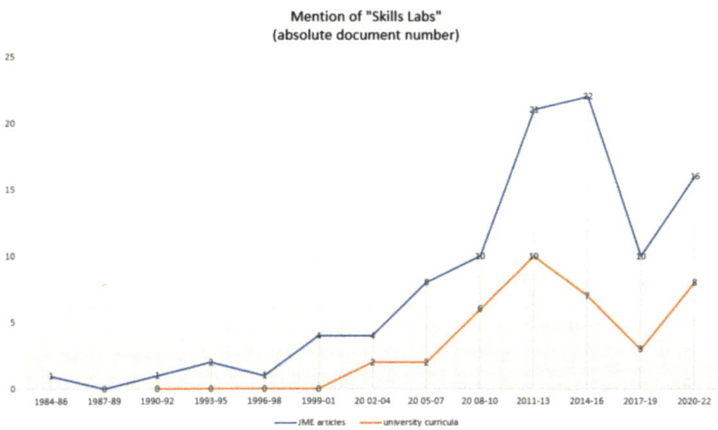

The lower graph shows a similar overall picture, although the total numbers are smaller. The same rapid rise and decline is evident, but the difference between the graphs is striking. JME articles start in the 1980s and show a steady mention of Skills Labs. In contrast, university curricula started mentioning Skills Labs in the 2000s. This shows that public discourse, as represented by the JME articles, are more likely to present Skills Labs as educational institutions. Conversely, individual universities begin to mention Skills Labs when setting up their own, which becomes possible after the 2002 examination order. This comparison is not a perfect representation of the rise of Skills Labs, but it does give a broad overview of the development and is consistent with the transformation overall.

Summary

In summary, the public arena analysis demonstrates the formal and informal negotiations of medical education within the public arena, with stakeholders outlining their agendas directly, such as GMA committees, and individuals taking stances or furthering research towards the establishment of didactics.

Such claims are relevant in two ways. First, they contribute to improving the subordinated position of medical education within the field of medicine. This is evident in the criticism directed at the Federal Ministry of Health and its institutions, such as the IMPP, and their reaction of producing the "Masterplan Medicine 2020". This is summarised in the section on SBT formalities. Second, medical educators working in Skills Labs who advocate for a competence-oriented, SBT-based education articulate claims against less reformed institutions within medical education. This becomes clear through the implementation and improvement of SBT commodities and unionising for didactic reforms, thereby making it a public issue.

The analysis of the public arena also makes clear that two critical junctures occurred within the considered timeframe of 1980s to 2020s; namely, the publication of the NKLM as internal shock and the COVID-19 pandemic as an external shock. Nonetheless, the NKLM's critical juncture and the overall development towards a competence-oriented education, as well as the implementation of SBT, can best be understood through the incremental change of constant agenda-setting of motivated educators. Further triangulation with these findings is conducted in Chapter 6, but prior to this, the third and final arena analysis must first be considered: the workplace arena.

5.3 Workplace Arena: Case Studies of SBT in North-Rhine Westphalia

In this final analysis, we move from the legal and public to the workplace arena, which Abbott describes as a "fuzzy reality" (Abbott, 1988, p. 66). In this professional arena, SBT educators are analysed in relation to their everyday work and self-reflection in the context of the broader medical education system. This third type of data describes SBT's structure and working conditions in Skills Labs at medical faculties in North Rhine-Westphalia, Germany. Interviews were conducted with personnel from the eight state medical study programmes (Aachen, Bielefeld, Bochum, Bonn, Düsseldorf, Essen, Köln and Münster) to find out more about the organisation of work within Skills Labs and the educators' stances towards SBT's implementation. In so doing, the variance from director to peer-to-peer teacher was taken into account in order to cover as many levels of personnel as possible. In addition to the interviews, it was also possible to visit almost all locations, with one exception. In this case, the visit was not possible due to the Skills Lab's being renovated. In all other cases, the researcher was either given a tour of the rooms by staff, took part in SBT lessons for participatory observation or, in one case, was even allowed to accompany an 'OSCE' examination process. Accordingly, the subchapter is divided into three sections: the organisation of the workplace (i.e., the Skills Labs' personnel structure); the practicalities of teaching (i.e., commodity management) and the definition and organisation of assessment of learning in Skills Labs (i.e., OSCEs).

Organisation of the Workplace

The personnel structure of the Skills Labs will first be explained in order to understand the organisation of work in Skills Labs as a workplace arena and with regard to the various functions and positions held by educators, but also how these can be understood in the context of the practical training system of both medical faculties and university hospitals. Furthermore, the intra- and interprofessional stratifications are just as important here as the self-image of the Skills Labs as a service unit for teaching in the faculties.

As shown in the figure 8 the Skills Lab's employees have various educational backgrounds, with some managerial, some regular employees and some specific roles. These include simulation person programmes and peer-to-peer teaching positions, where some are students and some are working professionals from different areas. The interviewees' responses can be condensed into four areas that personnel must address in order to operate a state-of-the-art Skills Lab: didactic concept, organisation, medical expertise, commodity management and simulation person programme. The latter two will be discussed under "Practical training". Manager recruitment is often important to the first two areas; however, many roles are combined into a single individual, particularly in managerial positions.

Consequently, there are 11 medical versus 7 non-medical professionals, including medical students in this sample. Five individuals cited supplementary training relevant to their roles, for example a Master of Medical Education (MME), group dynamics therapist and theatre pedagogy.

As mentioned previously, there are also various options for managing a Skills Lab. Having medical training is not necessarily a reason for recruitment, nor is experience in didactical training. However, if one or the other is not present, then the opposite needs to be supplemented, oftentimes creating multi-professional management (Transcript SkL03, Pos. 3, see Appendix 5; Transcript SkL12, Pos. 4–5, see Appendix 5).

Similarly, the personnel of Skills are Labs often diverse, including permanent employees, often for specified areas and peer-to-peer-educators who support teaching and organisation. Skills Labs can be entirely student-run in some cases. Being a non-medical educator (i.e., from social work, psychology or nursing science) may not initially suggest the possibility of working at a Skills Labs, however such backgrounds either lean towards the didactical concept-work (i.e., in SP programmes) or close to the medical profession, such as nursing. Similarly, peer-to-peer educators oftentimes consider their close proximity in proficiency as advantageous to creating a safe space in which students feel comfortable asking questions and taking ownership of their learning journey (Transcript SkL16.1+16.2, pos. 32, see Appendix 5).

Figure 8: Overview of the interviewees' educations

Conducted interviews	16
Interviewed individuals	18
∑ 16:28:27; Ø 01:01:47	
Social work	1
Sociology	2
Psychology	3
Theatre studies	2
Nursing science	2
Orthopaedic	1
Neurosurgery	1
Anaesthesia	4
Medicine unspecified	2

As explained in Chapter 3, the dichotomy of medical faculty and study hospitals within the medical education system results in not only staff from the Skills Lab itself, but also in clinical teachers to train and to assess within SBT. This relationship is particularly important and will be discussed in the next paragraph. However, the variety of clinical input becomes visible, with regard to the subdisciplines taught at the Skills Lab, through increasing mentions by interviewees: Tropical Medicine (mentioned 1), Pharmacology (1), Dermatology (1), Orthopaedics (1), Neurology (5), Psychology and Sociology (1), Surgery (8), Psychosomatics and Psychiatry (7), Urology (6), Gynaecology (8), Gastrology (1), Palliative Medicine (6), General and Internal Medicine (7), Paediatrics and Obstetrics (11), Cardiology (8), Otorhinolaryngology (4) and Anaesthesiology (8). This not only demonstrates the extent of the Skills Labs' offerings, but also illustrates the breadth of clinical fields represented within SBT. It would, therefore, be an inefficient use of resources to employ lecturers solely in the Skills Lab for each clinical area, as this would result in an unwieldy and inflexible institution, not to mention the necessary financial costs (Transcript SkL03, pos. 52–53, see Appendix 5). Instead, the two professional groups must cooperate in the delivery of education.

One reason brought forward by educators towards clinical teachers is the attempt to train students as well as possible and to kindle interest in specialist medicine so that they become better colleagues in the future (Transcript SkL15, pos. 118, see Appendix 5). Nonetheless, not all clinical lecturers are motivated to invest more time than necessary in education. This can be grounded in the subordination of medical education under medicine, and the lesser degree of abstract knowledge attributed to education. Instead, the educators bring forward social and cognitive claims and level them at didactical theory (Abbott, 1988, pp. 57–66). A better standing of SBT can commonly be traced back to a medical faculty or dean's office encouraging innovation in medical teaching (Transcript SkL09.1+09.2, pos. 88, see Appendix 5).

The presentation of the various organisational possibilities of Skills Labs has further demonstrated the considerable diversity of backgrounds among educators themselves, pointing toward this professional group's internal stratification. SKL07 outlines, for example, the rationale for recruiting a person with a medical background to a management position in a sub-unit of the Skills Lab, namely the SP programme (Transcript SkL07, pos. 61–73, see Appendix 5). Although SKL07 discusses the hierarchical order between medical and non-medical professionals, they also acknowledge the common language and knowledge that medical professionals possess that non-medical professionals cannot be expected to share. This raises the question of whether medical professionals from outside the department are unable or unwilling to communicate with individuals who are not accustomed to the clinical environment and professional norms. Nevertheless, SKL07 also underlines that there are other functions that they are unable to fulfil and that would benefit from additional personnel, such as psychological support in the event of triggering incidents for students or SPs. SKL07 is notably aware of this didactic necessity in the subsequent interview.

Other interviewees describe the discrepancies that they encountered between their clinical working and communication styles that they experienced in their professional lives and the didactic objectives pursued by the Skills Lab, especially among older colleagues (Transcript SkL08, pos. 71–75, see Appendix 5).

The issue here concerns not only different levels of speciality, but also different generations. Although the interviewees had a background in medicine, they have since moved into medical education particularly, in the Skills Lab specifically. This may result in them being shown less respect by clinicians. Another factor concerns the imbalance in teaching priorities in the Skills Lab environment. The primary objective is patient care in specialist departments within university hospitals. However, medical education is often neglected, due to pressure from the wider healthcare system. Consequently, medical educators must constantly justify the investment of both time and resources, emphasising their importance for the future of the profession. This is illustrated in the way that some interviewees find strong words: "I have to tell some professors that their teaching sucks [...]" (Transcript SkL15, pos. 8–14, see Appendix 5).

In conclusion, the interviews reveal different strategies for addressing the gap between the medical profession's clinical and educational sphere. Some interviewees argue with clinicians about their need to support the next generation of professionals while maintaining clear communication (SkL15). Some interviewees emphasise the importance of a shared language and attitude towards the healthcare system (SkL07). Others retreat and await a general shift in the outlook on education, hoping that the (seemingly less interested) generation will leave the work force (SkL08). In the latter case, this is combined with the hope that the next generation will work together more productively (Transcript SkL08, Pos. 85).

Explainable becomes this higher regard to the active clinician due to the fact that they represent the ideal that the medical education should prepare them for. "The professionals who receive the highest status from their peers are those who work in the most purely professional environments" (Abbott, 1988, p. 118). It is, therefore, unsurprising that students express a preference for hearing narratives drawn from clinicians' everyday experiences: "What they (clinical lecturers) are always good at is accompanying a case with their stories from the clinic. The students also like to hear that, their clinical expertise" (Transcript SkL03, Pos. 53, translation J.D.). This prompts the question what position can be attributed to Skills Labs within the institutional and professional framework of medical training at the faculties, specifically in terms of how far Skills Labs function as service units within faculties. This can be defined as a type of service unit for the purpose of teaching and possibly testing. "We offer the infrastructure [...] we offer the platform for diverse learning lessons" (transcript SKL05, pos. 42, translation J.D.). A variety of specialised medical professionals collaborate to develop practical teaching strategies in the Skills Lab. Employees in this institution, primarily serve as guides and intermediaries between the specialist clinics, in addition to potentially overseeing teaching units of their own (Transcript SkL01, pos. 23–24; Transcript SkL02, pos. 23–25, see both Appendix 5).

Employees must ensure the correct use of the rooms because the Skills Lab is a popular choice for teachers' practical lessons. This gives employees some influence over decisions (i.e., in design choices (Transcript SkL12, Pos. 104–106, see Appendix 5)). Managers and employees gain more influence as the Skills Lab's popularity grows. The historical developments vary from location to location. Large, modern Skills Labs are strongly integrated into the medical faculty (Transcript SkL06, pos. 32–33, see Appendix 5). This is because, as discussed previously, the Skills Lab's relevance as a safe space is not only a pedagogical concept, but also prevents errors in practice, which is important for patient safety. This emphasis is driven by humanitarian considerations as well as by market logic. The potential financial implications and reputational risks associated with medical malpractice or complications are significant, particularly for clinics and medical teams. Consequently, the necessity to invest effort and finances in medical training is reinforced by the potential consequences of such extreme cases (Transcript SkL15, pos. 203–204, see Appendix 5).

The Skills Lab personnel have a particular role to play that is, in line with Abbott's approach, negotiated, but that is firmly established in the Skills Lab development. The institutional composition directly impacts the workplace, requiring permanent employees and external teachers to agree upon organisation. This applies to fundamental decisions, such as establishing a Skills Lab, and day-to-day operations, including content and preparations for: the practical teaching.

Practical Teaching

According to Abbott (1988), professional skills consist of applying abstract knowledge from academic training to real-life situations through the three-step process of diagnosis, inference and treatment (see Chapter 2.1, 'Professions'). This is generally true of the medical profession too, but it raises the question of what constitutes professional practical application and what constitutes necessary skills. If these could be adequately defined, then the next question that follows is how they can be taught in practical training. The subchapter that follows addresses these issues, given that the Skills Labs are defined not only by the organisation of workers and their integration into the medical faculty structures, but also by the spatial arrangements and commodities used in SBT.

Many Skills Labs in North Rhine-Westphalia are in recently-constructed buildings. They are typically large and modern, with the extensive use of glass and concrete. All Skills Labs exceed the minimal definition of this research, namely that they require a minimum of three rooms. This phenomenon is also commented on in the public arena: "While at first, skills labs (sic) would be housed in buildings set for demolition on the edge of hospital grounds, in recognition of the effort spent, new designated buildings are now being built on prime real estate" (2016-V33-I4 (14), pp. 1–2). Other Skills Labs are in older buildings, some having been rebuilt, others having been demolished. New buildings are also under construction. This is common at universities, particularly medical departments, which often establish their own campuses and complex access routes. Sometimes these are on the outskirts and sometimes they can be found in the city centre. Skills Labs, as buildings, must be understood as having been designed solely for the acquisition of practical training.

Accordingly, interviewees deem specific rooms and spaces as essentials for a contemporary Skills Lab. These can be separated into specified and flexible rooms. Six distinct types of simulator rooms alone could be identified, beyond the minimum definition of simulation room, control room and debriefing room (Jaki et al., 2018, p. 23). Interviewees named full-scale, high-fidelity simulator rooms enabled by high-tech equipment, patient rooms resembling hospital or living spaces, emergency ambulances simulating first aid services, operating rooms (which are often similar to high-fidelity simulator rooms), but which are also equipped with operating lights and sterile access routes, shock rooms resembling clinical emergency rooms and, finally, there are doctor's rooms equipped with a desk, chair and a stretcher. However, not all Skills Labs have enough space to house all of the rooms mentioned; instead, they opt for flexible rooms that allow trainers to rearrange the space for different training sessions. In this regard, space can be understood as a scarce resource in Skills Labs, often necessitating moves to larger facilities (Transcript SkL02, Pos. 49, see Appendix 5).

The lack of space and layout represents a significant challenge for those designing and conducting training sessions with simulators and with simulated patients. Ensuring sufficient space for movement is essential to enable students to either interact effectively with the study object or to facilitate a focused conversation. Not only does this include the simulation rooms, but it also includes the control and debriefing rooms. The control rooms are often equipped with double-sided mirrors or video transmissions, but both can produce problems. Double-sided mirrors ask for a very specific layout of the building, something which is not ideal in all cases (Transcript SkL10, Pos. 38–39, see Appendix 5). Video transmission can also produce problems with technology and can distance the observers from the simulation activity observed.

Moreover, in a small number of cases, the requirement of three rooms per part of the simulation has not been met. Instead, the observing students, the simulating students and the discussion of training afterwards are placed in a single room. This has the potential to generate noise and distraction, which can disrupt the simulation's efficacy. Nevertheless, most of the Skills Labs observed were found to contain a greater number and variety of spaces than would be expected for the minimal definition.

This extends beyond the rooms used for the immediate SBT and encompasses the Skills Labs' organisational processes more generally. These include make-up rooms for simulated patients, office space for educators and extensive storage facilities (Transcript SkL14, Pos. 59, see Appendix 5). The extensive list of the Skills Labs' simulators, and their diverse portfolio of specialised and flexible rooms used by various disciplines for teaching, demonstrates the need for storage space when other lessons are in progress. One interviewee suggests the ideal of as much storage space as teaching space (see transcript SkL15, pos. 170, see Appendix 5), while others advise devoting 20–30% of the building for storage (Jaki et al., 2018, p. 46). However, the necessity for well-organised storage with clear labels and records of materials in use or undergoing repair was emphasised repeatedly in interviews and during visits to the storage rooms and back doors. In conclusion, space is vital for the analysis of Skills Labs. This is because of its role in the growth of these institutions. It is also important in terms of design and intention. As with any institution, the rule-making and intention of design is often not congruent with the actual use and can have unintended consequences (Thelen, 1999, p. 382). This is especially true when space is missing for a specific need that only becomes clear in practice, such as offering students learning spaces or planning for enough storage room. Jaki et al. (2018) offer a list of factors to consider when establishing a Skills Lab; in addition to the minimal version quoted above, these include: location, security, entrance/waiting area, simulation rooms, control/observation rooms, debriefing room, changing rooms, storage rooms, office and common rooms, possibly a café-like area (Jaki et al., 2018, p. 40). As outlined previously, the spatial composition works best when it is closely aligned with the commodities used. The term commodity originates in Abbott's theory of professions (1988) and refers to objects that embody a profession abstract knowledge (see Chapter 2.1.).

The following sections describes the commodities used in SBT. Firstly, simulators are material representations of anatomy and procedures and are usually made of plastic. The second category of commodities are simulated patients, actors and actresses playing sick patients or their relatives. Thirdly, there are SBT-related commodities, which exist in these trainings, but which are not involved either exclusively or in every case.

Simulators

As described in Chapter 3 on SBT, simulators can be differentiated into three categories: simple models, simple dummies and high-fidelity manikins. The following examples from the Skills Lab highlight the usefulness of separating the simulators according to their functions and the extent to which educators themselves are critical of the simulator's abilities. One interviewee mentions an argument often used to promote SBT: you cannot harm a dummy or model (Transcript SkL06, pos. 83, Appendix 5). Accordingly, students can also practise their skills on simulators. However, the "own laws" of applying practical skills rely on doing the physical steps yourself. This is more difficult than observing a complex procedure, as students found with simulators. In one instance, students had to rescue an unconscious person and perform CPR, defibrillate, evaluate heart signals and manage an airway. Despite theory having been made clear, students frequently asked for clarification regarding the defibrillator, the manikin response to the simulated shock and the button to press prior to entering the simulation. In an interview, the educator in charge confirmed that this is typical of SBTs involving high-fidelity manikins (Transcript SkL13, pos. 50–51, see Appendix 5). The ability to perform the processes required independently is a fundamental aspect of SBT learning, whether the training aid is a high-fidelity manikin, a simple dummy or a model. The objective is always to gain hands-on experience of the objects and to internalise the processes through experience. Even still, it may not always be a perfect representation of the procedure itself (see Transcript SkL12, Pos. 114; Transcript SkL01, Pos. 72–73, both Appendix 5).

The simulators exemplify SBT's fundamental concepts: repetitive, safe practice, specific requirements pertaining to the scenario and the corresponding clinical knowledge. They also demonstrate predictable procedures, at least on the part of the educators, when teaching lessons to students. Furthermore, they implement study objects that are sometimes controlled by complex technology, but that can also be comparatively simple objects or representations of human anatomy. All of these concepts are combined within one institution, in dedicated rooms and with rapid temporal changes between the simulators. SKL08 provides an illustration of this concept by comparing the various weeks and focal points addressed by their Skills Lab (Transcript SkL08, Pos. 23, see Appendix 5).

However, simulators cannot cover all aspects of medical practice, especially when it comes to communicative procedures. This is where simulated patients fulfil another function of commodities in SBT.

Simulated Patients as Educational Tools

Having described the simulators in detail, what follows examines interesting SPs. Many Skills Labs have many simulators that are popular with students. However, as SKL03 makes clear (see Transcript SKL03, Pos. 26–27 in Appendix 5), they only represent part of professional practice. Simulators can never replicate human reactions or appearances and they cannot respond to questions. Therefore, some educators argue that SPs are superior. SPs are most commonly used for communication training. Here, one student can act out a scenario while others observe via a double-sided mirror, video transmission or can even occupy the same room. Students can take notes and provide feedback without being haptically involved initially. The SP acts within the simulation and can also provide feedback afterwards.

Additionally, there are props that serve a specific purpose, such as a gown to represent a professional role or gloves to indicate a clean environment. The SPs are often regarded as educational instruments that are contingent on lesson focus (Transcript SkL10, pos. 7, see Appendix 5). However, there are also other perspectives that advocate greater involvement of the SPs in the educational concepts (Transcript SkL09.1+09.2, pos. 113, see Appendix 5). The requirements for becoming an SP vary. Individuals fulfilling the role of SP usually have a background in acting (see transcripts SkL09.1+09.2, Pos. 55–57; SkL07, Pos. 22–23). They must also undergo training in medical education, particularly in terms of constructive feedback and exit strategies to avoid psychological distress (Transcript SkL07, Pos. 55). Ensuring the psychological wellbeing of SPs and students in triggering situations is, therefore, also a task for Skills Lab staff.

In comparison to simulators, the SPs offer a unique combination of practical skill and communication training. In the example of taking blood, the practical skill can be well-trained on a simple dummy, but the dummy requires no communication, such as reassurance if the insertion hurts or if the first attempt at piercing is unsuccessful (Transcript SkL05, Pos. 78, See Appendix 5). Hybrids are increasingly being used to combine the advantages of simulators and SPs (Brown & Tortorella, 2020). Hybrids are simple models that are attached to an SP by extension, which not only covers practical skills, but also correct movements and the continuation of communication. As with the SP in isolation, real individuals of different sizes and with different reactions and possibly different emotional role scripts can be represented, while the student actually performs the practical task at hand. However, while some interviewees welcome such additions to SPs (Transcript SkL07, pos. 39–41, See Appendix 5), others argue against them in terms of lower fidelity (Transcript SkL05, pos. 80–82, See Appendix 5).

Lastly, the same factor that makes SPs more advantageous than simulators, namely that students must interact with real people, and this can also be a disadvantage if the SP makes it too easy for students. A study from 2015 evaluated the SP interaction at a Skills Lab and found that some SPs tend to simplify their reactions over the course of several semesters, while others remain stable in their communication (Kliche, 2015, p. 104), something which SkL14 also observes in their everyday work (Transcript SkL14, Pos. 25, see Appendix 5). In this respect, the SPs can be understood as commodities, according to Abbott (1988), by offering the same performance to different students over and over again, because they are only primarily grasped by the students at the moment of the simulation in terms of the information necessary for medical practice.

The Verge

This concluding section on commodities in SBT comprises a list of various objects that verge on SBT. To wit, they are either related to or used in SBT, but that are not exclusive thereto.

Consumables and Medical Instruments

It can be reasonably deduced that consumables and medical instruments, which are both part of and used in normal medical practice, are also used in SBT. In most cases, the specific materials are deliberately utilised with the objective of keeping the conditions as realistic as possible. It is, therefore, crucial for students to learn how to use these items correctly, by putting on sterile gloves to practise hygiene and to gain familiarity with the routine of conducting a physical examination, for instance (Underman, 2022, p. 188). While these may not be considered to be SBT commodities typically, they are integral to the workplace and must be sourced, organised and made available for training (Transcript SkL12, pos. 64–65, translation J.D.).

Self-Made Sensory Analogies

SkL03 illustrates a transformative context because Skills Labs lacked the same level of funding as other institutions in their initial stages of development and, therefore, had to be more resourceful in their approach. This included the use of improvised methods and excluded simulations with high-fidelity manikins, which were not within their financial reach. However, the statement "you can simulate with a potato sack" (Transcript SkL03, Pos. 21, See Appendix 5) emphasises that SBT's success is not determined by the technology used, but rather by the educational input. In reference to the work of Harris and Harder, this can be understood as a sensory analogy, which is not directly a simulation with the pre-produced objects like simple models, dummies or high-fidelity manikins, but is rather a sensory learning with everyday objects (Harris & Harder, 2022, p. 51).

VR, AR and Immersive Projection Rooms

As shown in the interview with SkL03 (Transcript SkL03, pos. 27–29, see Appendix 5), students find the field of high technology attractive, especially when it is constructed digitally. SkL06 describes this with their augmented reality room, the sim arena in which different scenarios can be projected. One example is an outdoor scenario, while another is a projected surgical environment (Transcript SkL06, Pos. 27, 101, See Appendix 5). However, while the interviewees describe students' enthusiasm and the educators also see potential in the technology, the technical requirements remain critical. VR, AR and other solutions are expensive commodities for Skills Labs. Not only may spatial changes be necessary for their use, but the management of these commodities also needs a high level of expertise. As with high-fidelity manikins, trained personnel need to familiarise themselves with the technology and should be in a position to repair it if problems arise. Even more significantly, the learning objectives associated with VR or AR applications tend to be small (Transcript SkL12, Pos. 112, see Appendix 5).

In conclusion, the commodities on the verge of SBT are diverse and not exclusive to SBT, but they often enhance the fidelity of the simulation scenario or expand upon the notion of pre-existing commodities towards other solutions, both consisting of everyday items or digital solutions. The following section provides further explanation of the concept of fidelity and management of commodities, which has been mentioned repeatedly.

Fidelity, Management and Repair of Commodities

SBT's transfer to a clinical setting is a possibility that has been demonstrated by studies, particularly in the context of surgical procedures where students are able to demonstrate psychomotor remembrance by comparing simulator and patient outcomes (Agha & Fowler, 2015). Therefore, the objective of educators is to achieve a high level of fidelity to the clinical environment, without the intention of reconstructing it in its entirety (Jaki et al., 2018). The fidelity argument can be applied in three ways. Firstly, authentic materials can be used to closely represent the clinical setting. Secondly, simulators, SPs or VR/AR can be employed, even though they are not representations of 'real' patients. Rather, they showcase specific aspects of clinical cases, making them commodities of simulation-based education. Thirdly, inauthentic materials can be used to provide a more authentic sensory experience of clinical circumstances. Fidelity is a key concern in the SBT workplace arena for those working in the Skills Labs, given that they must maintain it at the highest possible level. As SkL05 stated in the context of hybrids, an implausible model can disrupt the overall situation by drawing attention to its simulated reality (Transcript SkL05, Pos. 80–82). Accordingly, there are no fixed limits to when fidelity is or is not reached; it can only be achieved in a gradual manner.

The SKL01 transcript (see Appendix 5, pos. 50–51) reiterates the objective of fidelity in SBT: to facilitate the transfer of knowledge from SBT to clinical procedures. However, some educators clarify that a high-fidelity simulator's superiority does not inherently guarantee its effectiveness. The value of such a simulator depends on the specific skills and procedures practised in the training context. SKL12 gives an example of this with blood sampling, where a high-fidelity manikin is unnecessary and expensive for this minor procedure; simple models of individual arms are enough as their parts are simpler to replace (Transcript SKL12, pos. 75).

The level of fidelity required for a simulation room must be considered alongside the fidelity of the commodities. Specific training rooms can be established within Skills Labs to represent authentic settings, such as a doctor's surgery or a patient's room. Studies also indicate that simulation situations outside the clinical setting benefit medical education (Backlund et al., 2018). However, the required level of detail for furnishings can be flexible because it is mostly the commodity (i.e., an SP) that represents the training situation (Transcript SkL14, pos. 67; Transcript SkL07, pos. 30–32, for both see Appendix 5).

Similarly, high-fidelity simulators' advanced technology can jeopardise their fidelity if an educator is unable to fix technical difficulties (Transcript SkL14, pos. 73, see Appendix 5). This raises the question of how educators define and improve the fidelity of SBT. As noted with sensory analogies, this requires imagination, experience and expertise, particularly medical expertise, to create a specific clinical situation. SKL15 (Transcript SkL15, pos. 72–73, see Appendix 5) describes how the simulation has to 'feel right'. The construction of the situation and this sense of fidelity reproduces the clinical environment, reducing illness and the human body to (seemingly) only those factors that are medically relevant. However, this has been criticised (Nott & Harris, 2023, p. 217). It is, therefore, analytically useful to consider fidelity, linked to spatiality and commodities, as a carrier the profession's abstract medical knowledge – just like the commodities themselves. SBT's orientation towards fidelity produces and reproduces professional knowledge that is neither neutral nor given simply (Abbott, 1991).

Both the management and, potentially, the repair of commodities is of central importance. However, it is also bound to the formal institutional circumstances. SKL08 outlines the procedure at German universities of requesting equipment as being so time-consuming that, by the time equipment arrives, newer models are available on the market. (Transcript SKL08, Pos. 40–47, see Appendix 5). This shows how complex equipping a Skills Lab is, even if the time estimate is probably exaggerated. Ongoing maintenance also needs to be considered. SKL09 commented on the 'hype' around high-fidelity manikins because they are versatile, but often malfunction. They, therefore, prefer 'the simple things that don't break'. (Transcript SKL05, Pos. 91, translation J.D.; transcript SKL13, Pos. 5, see Appendix 5). In the long run, sending in and repairing devices disrupts the simulation process, as they are not available during that time. Therefore, other tinkering sometimes takes place. In one case I observed the use of a high-fidelity manikin and an additional device to improvise and to avoid the need for constant repairs (Transcript SkL13, pos. 6–9, see Appendix 5). Accordingly, SKL14 described the market structures as too narrow and, therefore, simulators were less reliable (Transcript SkL14, pos. 75–76, see Appendix 5). The hypothesis proposed by SKL14 cannot be either verified or falsified within this study's confines, given that it would require the involvement of additional research with manufacturers themselves.

Nevertheless, the examples from the workplace demonstrate the considerable time investment required for the management and repair of commodities, as well as the execution of practical teaching overall, in Skills Labs. The dependence described once again underlines the stickiness of institutions, in this case in the management of commodities with regard to bureaucratic procedures of purchase and repairs. However, it is not just the commodities themselves, but also SBT's learning goals that structure the practical teaching in Skills Labs. The advantages that the educators see in SBT compared to other forms of (practical) teaching need to be considered.

Educators almost universally emphasise that SBT should not be the only form of teaching, but that it has advantages over other formats as practical training (Transcript SkL02, Pos. 4, see Appendix 5). In general, Skills Labs in medical pre-graduate education are mostly associated with the training of basic skills. A large number of basic skills are also mentioned with the following frequency in the interviews with Skills Lab staff: Medication 1, imaging (CT, MRI, X-ray) 4, percussion 2, hygiene 4, puncture + access 35, ECG 10, emergency medicine 31, suturing techniques 6, sonography 16, palpation/haptic examination 16 and auscultation 14 times. These skills represent those practical behaviours that every medical professional must be able to perform, such as interpreting a variety of diagnostic data, but also applying various examination techniques. However, this also requires the specific commodity to be aligned with the respective learning objective, which in turn is a question about the curricula as the legal basis for the training, as opposed to the workplace arena in which actual implementation is carried out (see Chapter 6.2 Triangulation: legal versus workplace arena). This is why the educators in the Skills Labs are well aware of what their training courses can offer that other forms cannot, which SKL09.1, for example, negotiates under the concept of constructive alignment (i.e., the matching of learning objectives and examination objectives (Transcript SkL09.1+09.2, Pos.100, see Appendix 5)). SKL12 also recalls their own experience with bedside teaching as insufficient and inappropriate towards real patients and being "looked at like a tiger in the zoo" (Transcript SkL12, Pos. 140, see Appendix 5).

Abbott would consider this problem to be one of optimum abstraction (i.e., how and how much theoretical knowledge in comparison to practical skills are being trained with (Abbott, 1988, pp. 105–106)). This encapsulates the challenges inherent to clinical training. Ideally, students should observe the presentation of specific diseases, experience the impact of certain conditions and practise applying their skills. However, the clinical environment does not always offer this, so the Skills Lab provides an alternative learning experience. For SBT educators, this involves overcoming the challenges associated with implementing SBT, particularly with regard to fidelity and commodity management. Furthermore, after teaching practical skills, educators must also assess students in practical exams, which will be discussed below.

Assessment of Learning

As outlined previously, practical teaching can never reflect the full clinical experience of the medical profession. This is consistent with Abbott's remarks about the difference between abstract professional knowledge and its practical application: "In reality, this perfected abstract knowledge system exists only in the world of professional textbooks" (Abbott, 1988, p. 56). The issue with assessment, however, is that it is the touchstone of the professional identity of individual students. Each student must demonstrate their theoretical knowledge and practical skills in test situations. The Skills Labs are involved in the development of such practical examinations in various ways, with objective structural clinical examinations, short OSCEs, and entrustable professional activities, short EPAs, for example, both of which are rooted in competence-oriented education.

Competence is hard to define and to assess. This is partly because medical education is oriented towards the clinical encounter, which is versatile, and assessment should reflect it. Conventional assessment is directed towards knowledge that is internalised, meaning an assessment *of* learning, while a competence-oriented assessment would need to show gaps in knowledge and ability, meaning an assessment *for* learning (Fabry, 2022, pp. 296–297). It is challenging to create assessment concepts that give educators feedback on learners' proficiency and that provide learners with a protective space in which to develop their professional abilities. Two concepts that aim for assessment for learning are OSCEs and EPAs.

OSCEs, an acronym for objective structural clinical examination was introduced previously, in both the legal and public arenas. *EPAs, an acronym for entrustable professional activities*, conversely, is a concept first published in 2013. Its procedure is based on the same idea as supervision: The feedback from a master that would trust the novice with a given task (Cate, 2013). The usage of EPAs forces a curriculum to define skills that are more or less critical to gaining the status of a professional as well as aspects of trustworthiness that the educator can fall back onto, while leaving room for the latent aspects of education that the competence-orientation require.

When asked about participation in examinations formats, SKL02 explains whether or how they are involved in examinations that it varies greatly and that it depends on the Skills Lab's respective site (Transcript SkL02, pos. 81). Accordingly, it is worth taking a comparative look at the sites in North Rhine-Westphalia. OSCEs were discussed in the workplace arena in seven out of eight cities and were conducted as a regular examination format in six cities (Aachen, Bielefeld, Bonn, Düsseldorf, Essen, Cologne). At one site, the interviewees were not involved in examinations and were, therefore, unable to provide any information (Bochum). In another case, the institutional composition is responsible for the fact that no OSCEs (in the classic sense) are carried out. In this case, the Skills Lab is divided into two parts, a training centre and an assessment centre (Münster).

The preparation of OSCEs is a work-intensive process for educators in Skills Labs. It entails numerous, communicational feedback loops between clinical specialties that are assessed, sometimes resulting in a disconnect between the needs of the Skills Labs personnel and the preparation of clinical teachers. Furthermore, each OSCE station requires the thorough preparation of commodities to fulfil the tight time schedule. Nonetheless, OSCEs are a commonly used and prominent tool for practical assessment. According to recent examination order drafts, the OSCE could even have become part of the state examinations. However, its deletion was welcomed by educators. This reaction may be unexpected, given that the introduction of OSCEs, which are based exclusively on simulations, would typically accord greater significance to Skills Labs within the context of medical training and in the respective medical faculties. Instead, educators primarily expressed concern about the feasibility of their workloads (Transcript SkL08, Pos. 11; Transcript SkL07, pos. 44–45, for both see Appendix 5).

Similar explanations can be provided for the differences in the use of OSCEs at various universities with different commodities and funding (Transcript, SkL09.1+09.2, Pos. 78–80, see Appendix 5). Nevertheless, the fundamental issue with all practical exams remains the dilemma mentioned above. It is an imperative that exams are comparable for candidates, while simultaneously allowing for sufficient variability so as to demonstrate the adaptability of students (i.e., clinical skills). As per Abbott's theory, this can be defined as the connection between diagnosis and treatment (1988). An additional assessment tool has been developed with the purpose of increasing the variability that is permitted.

Like OSCEs, EPAs are based on demonstrating competence in practical procedures. These assessments evaluate students' confidence in a way that is different from simple checklists. Instead, examiners indicate their trust in students' abilities (Transcript SkL06, Pos. 59, see Appendix 5). The principle of trust is important here because it represents an underlying category that avoids the problem of excessive standardisation. At the same time, as the quote describes, clear guidelines are needed to determine what constitutes a particular area of expertise and needs clinical examiners who are willing and able to engage in this assessment process, even over a longer period. This supervisory feedback provides students with realistic insights into their self-perception of their ability (Transcript SkL05, Pos. 19, see Appendix 5). Personnel shortages persist, however and much like in actual patient care, thereby making EPAs a more aspirational model of medical education than a tangible point of reform at present. Conversely, competency-orientation is inextricably linked to SBT, thereby making it a prominent topic of discussion in the workplace arena.

Competence of Students

There are two ways in which competence, as professional behaviour and proficiency after Abbott, need to be considered before summarising the workplace arena's findings. First, the students themselves are subject to educators' professional practice. Second, the students are meant to have become professionals themselves after their studies. Therefore, the competences expected of students (as future professionals) and the competences of educators (as professionals) within the medical educational institutions that disseminate their practice and knowledge requires attention. Some interviewees link the discussion around medical education to the wider educational culture in Germany. SKL09.1 criticises the reproduction-based 'bulimic learning' approach to the German education system, which clashes with the transfer of knowledge-based and competence-oriented approach. They argue that this weakens the position of reformist educators (Transcript SKL09.1+09.2, Pos. 76). Similarly, SKL16.1 argues that practical competencies are more desirable for students and criticises the excessive theoretical input of medical studies, which they claim fosters 'nerds' (German: Fachidiot) (Transcript SkL16.1+16.2, Pos. 200). SBT educational goals are oriented towards competencies, unlike the conventional approach. This entails a shift away from highly specialised disciplines and towards organ-oriented issues, in combination with distinct competencies. Three categories of competencies can be distinguished: communicative competency, clinical or hands-on competency and, to a lesser extent, moral competence or the professional identity's values. The following sections offer a synopsis of each of these core competencies.

Communicative competence is primarily aimed at taking a medical history (i.e., addressing the patient and their suspected diagnosis appropriately, sometimes under time pressure). A common medical exercise is called "breaking bad news", where SP as patients are informed of their diagnosis and are then reassured by the students as doctors; this requires both significant medical and communicative expertise. The advantage of SPs is that they do not suffer emotional damage if a communication by a student was not appropriate in the training situation (Transcript SkL10, Pos. 19–21, see Appendix 5). There are various models, such as 'NURSE' or 'SPIKES', to help students to structure difficult communicative situations and that cannot be explained further here for reasons of space (transcript SkL09.1+09.2, Pos. 49). Questioning techniques, such as motivational interviewing to motivate lifestyle changes (Transcript SkL07, pos. 16–17) or the 'anamnesis house' as a metaphor for different areas of the patient's life (Transcript SkL16.1+16.2, pos. 69–71), can also assist students with taking control of a conversation, thereby creating an alliance with the patient for further treatment.

Hands-on or practical clinical competence is difficult to define because it covers a variety of professional behaviours. In its narrowest sense, it encompasses the fundamental abilities required for basic procedures, which all medical practitioners should possess. Several interviewees mentioned a greater focus on hands-on skills and a reduction in the emphasis on theoretical knowledge, with the aim of better preparing students for the clinical environment (Transcript SkL08, pos. 66–69, see Appendix 5). However, clinical competence entails not only standardised procedures, but also adaptability to unexpected circumstances. Consequently, some SBT programmes emphasise small inconsistencies across simulation settings (see Transcript SkL12, pos. 112–114).

Moral competence and the values associated with a professional identity represent a latent competence, which makes them more challenging to train and to assess within the context of medical education (Transcript SkL05, Pos. 19, see Appendix 5). Nevertheless, they are a significant aspect of the societal role that doctors occupy as part of a country's healthcare system. Doctors assume responsibility for their patients' wellbeing and may inflict harm if they practice incorrectly (Transcript SkL02, pos. 6–9, see Appendix 5). It is frequently proposed that students' moral competence should be regarded as an acknowledgement of their responsibility and as an understanding of their individual limits, similar to the concept of EPAs. A close relationship between student and educator is, therefore, also necessary, since values are most effectively transmitted by serving as a role model (Transcript SkL06, Pos. 89–90).

Nevertheless, a particular point of criticism is often raised by SBT educators in the context of medical education, and this is a critique that is also applicable to SBT itself: The division of knowledge areas, the distinction drawn between hands-on and communicative skills in this case. It finds an analogy in the spatial arrangement of one of the Skills Labs that I visited. In this instance, the Skills Lab was entirely segregated into two distinct sections: one dedicated to communication in one of the building's wings and the other, for hands-on skills, was in the other wing. However, Abbott himself relativises this point of criticism by stating that diagnosing and treating can be formulaic, while inference requires a deeper, contextualised consideration and, thus, entails true professionalism (Abbott, 1988, p. 48). In the context of training scenarios, it is permissible to minimise reality to focus learning on specific aspects. However, the selection of the relevant aspects for the competence development of students requires certain competencies on the educators' part.

Competence of Educators

It is noteworthy that authors who describe how SBT should or could be integrated into university curricula do so in terms of the educational objectives that they seek to achieve, which are didactically competence goals. This observation suggests that SBT and competence-orientation are inevitably intertwined (Ahlers, 2018). For the SBT workplace arena, this means addressing the medical system's demanding conditions, while simultaneously promoting practical teaching, for which SBT provides a solution (Transcript SkL02, pos. 101–103; Transcript SkL11, pos. 10–11, for both see Appendix 5). With this argumentation it also becomes apparent why 'teaching the teacher' is an important concept in medical education. In Germany, the Master of Medical Education, MME, is the main programme for medical educators. The MME is associated with the MFT (Medizinischer Fakultätentag (MFT) Deutschland,

2009) and was established in 2004 (Medizinischer Fakultätentag (MFT) Deutschland, 2013, p. 11).

Furthermore, it is important for all teachers in the Skills Lab to be familiar with the didactic concepts, as this is the only way to design courses that are in line with the learning objectives. As described, some educators note that whether and how involved educators are in further training for didactics can be a generational issue. For example, SKL08 describes that the "the younger ones, who are just about to habilitate, they also make these LAMA courses and such. They are engaged and care. But yes, we also fight a lost cause sometimes [Ger.: Kampf gegen Windmühlen], because we can tell them (older or less interested individuals) five times which ways are not acceptable. (They respond) 'Yes, it is, it has always been like that'" (Transcript SkL08, Pos. 81, translation J.D.). The mention of LAMA courses in the quote refers to the Academy for Medical Education in North Rhine Westphalia (Ger.: „Landesakademie für Medizinische Ausbildung Nordrhein-Westfalen", short LAMA), which is a cooperation of all of North Rhine-Westphalia's medical faculties and issues certificates for additional training in medical education (Geschäftsstelle Heinrich-Heine-Universität Düsseldorf, 2024).

The educators emphasise that trainers in SBT, and in all areas of medical education require, certain skills to fulfil an educational purpose, treating students as future professionals and motivating them to fulfil their role independently and dutifully (Transcript SkL15, pos. 119–120, see Appendix 5). In the context of the broader advancement of medical education, the roles played by the MFT and GMA are particularly significant. Educators oftentimes refer to specific networking activities within the SBT community to help facilitate a high standard of education and to exchange communications about developments (Transcript SkL15, pos. 192, see Appendix 5).

Several important committees within the GMA are mentioned, in which the educators are involved and which they see as central to networking. The most central committees include the Committee for Practical Skills and the Committee for Simulation Patients, but the Committee for Assessment, the Committee for Communicative and Social Competence or the Committee for Interprofessional Education are also mentioned (Transcript SkL15, Pos. 196; Transcript SkL14, Pos. 90; Transcript SkL09.1+09.2, Pos. 61; Transcript SkL01, Pos. 82; Transcript SkL12, Pos. 118; Transcript SkL06, Pos. 97; Transcript SkL03, Pos. 31).

Some of the interviewees describe themselves as having led, or currently leading, developments within the committees mentioned. Some interviewees confirm that even if they do not participate in committees themselves, they observe and use the results of such committees in their daily work, to certify to their home faculties that certain educational standards are up to date for example (Transcript SkL05, Pos. 73). Locally, other networking functions with regard to SBT include the Didactic Network North Rhine-Westphalia with its Simulated Patient's Network, in which several of North Rhine-Westphalia's universities participate and exchange information about their SP programmes (Transcript SkL03, Pos. 31; Transcript SkL06, Pos. 97). Internationally, there is the Skills Labs Symposium, in which leaders and educators from Skills Labs from at least all German-speaking countries meet at an annual conference (Transcript SkL02, Pos. 81–83; Transcript SkL13, Pos. 37; Transcript SkL12, Pos. 35; Transcript SkL08, Pos. 77; Transcript SkL01, Pos. 106). Finally, it should be noted that the development of new iterations of the NKLM is an ongoing process. Consequently, numerous educators within the Skills Lab sector are engaged in subdivisions of the NKLM and work towards the delineation of communicative and practical competencies (Transcript SkL03, Pos. 43).

Summary

In conclusion, the establishment of a professional community in the workplace arena and the formulation of standards or a code of ethics in medical education are driven by several factors. Primarily, they serve to reinforce the inner circle's identity, unify the understanding of exemplary professional conduct, in this case efficient SBT, and exclude those who do not adhere to these aspirations (Abbott, 1988, p. 5). It is evident that the Skills Labs provide a multitude of analytical value with regard to the understanding of professionalisation and the institutionalisation of SBT, as was articulated in the preceding description about the workplace arena. The organisation of the workplace, in terms of personnel, faculty composition and social stratification can be interpreted as a manifestation of institutional forces and stakeholders that either accelerate or dampen Skills Labs' institutional development. The characterisation of the Skills Labs, as service units, particularly in conjunction with the reflections on space and commodities, and the extensive variety of considerations pertaining to management and repair, demonstrate the specific knowledge and, consequently, the power that the Skills Lab staff can exert within the medical faculties and MES. Finally, the assessment of practical training is often constructed and bound to the Skills Lab facilities, with the aim of reflecting the needs and competencies of a changed attitude regarding how students should be taught and prepared for their future professional roles. Therefore, the potential for transformation yielded by the workplace arena should not be underestimated. Developments, such as novel uses of commodities, can enhance professionalisation if the professional networks described follow such examples, forcing even clinical teachers (who are not focused on SBT) to redirect their teaching and assessment: "This pattern of power assumes, correctly, that jurisdictional invasion generally begins in the workplace, then moves to the public mind, and then into the law" (Abbott, 1988, p. 139).

5.4 Open questions towards triangulation

Following the discussion of the three arenas separately, the insights obtained and the remaining questions need to be summarised briefly. The rise of medical education's professionalisation towards SBT as well as the formal and informal institutional forces shaping this transformation are represented differently in the three arenas.

In the legal arena, the curricula and regulations are regarded as the formal guidance to medical studies in Germany. However, significant differences are found between the nationally-binding examination order and the individual study programmes, especially towards the mentioning of SBT. These documents are formalised to the degree of paragraphs and appendices, given that they need to be legally binding. The informal is rendered invisible within this sort of text because this analysis would necessitate insight into the argumentation and decision-making process that led to the formulation of a particular paragraph, whether a part of the text was omitted or added or whether there has been debate about the content in the first place. The comparison can show changes down through the years and these are valuable in understanding SBT's formal side (i.e., how it is defined, if mentioned) but leave little indication about what institutional resistance, if any, had to be overcome for it to be included.

By contrast, the public arena is characterised by the presence of both formal and informal indications. The JME and its publisher, the GMA, are regarded with a certain degree of formality. This is because both institutions have both a set of regulations and a formalised way of publishing. In addition, the GMA's various subsections have given themselves an agenda within the broader framework. Nevertheless, the publications' contents may comprise formal information pertaining to networking activities, the organisation of events or to the description of Skills Labs as institutions (such as concerning their personnel, funding, commodities, and so forth). However, when authors either advocate for or oppose particular educational or institutional advancements, they are also articulating their positions, in part as individuals and in part as representatives of institutions. It is imperative to note that the transition from the formal to the informal understanding of these publications, or parts thereof, is gradual. These publications do not represent a dichotomy, instead exhibiting a shift towards one perspective or another. It is evident that these phenomena do indeed serve a role in the establishment of agendas, and their presence can be indicative of the existence of informal agreements (see Chapter 2.1). One may consider the establishment of a new network activity in order to illustrate this point. It could be argued that this indicates that communication and knowledge about other likeminded parties must have been anticipated, if not ensured beforehand.

Lastly, the workplace arena also represents both formal and informal aspects. The establishment of a Skills Lab requires specific funding from the medical department's parent institution. The personnel are bound by a contract and the position is justified and provides a specific reason. The space and commodities are connected to contracts and formal agreements with the Land and manufacturers, for example. A Skills Lab may even have a formalised self-description in the form of a booklet or website representing its educational agenda. However, although the personnel are hired for a specified reason, their role or stance towards the institution may change. They can be more or less involved in improving SBT and its formalised state by, for example, publishing statements or networking with other institutions. While the training of students is formally articulated in the curricula, the concrete details of teaching and assessment are informal and everyday decisions taken by members of institutions can transform its regime over time, as assumed by HI (see Chapter 2.2).

The following chapter triangulates the three arenas and, in accordance with the assumed interdependence of professional arenas, follows Abbott (1988) and examines the incremental changes identifiable through formal and informal rules after HI. This enables a more accurate understanding of the transformation and allows for the inconsistencies of the various data points to be weighed up against each other. Consequently, no arena remains isolated in its narrative and logic. Instead, the various stakeholders and levels of social forces can be analysed more fully.

6. Triangulation of jurisdictional arenas

This chapter will expand upon the discussion of the finding outlined by triangulating the three arenas of professional jurisdiction. This strategy's objective is to deepen the analysis by comparing and revealing supplementary data points. This includes all stakeholders within the Medical Educational System in Germany that were found inductively and deductively through the content analysis conducted and that must, therefore, be considered for SBT's professionalisation (see also Chapter 3.1). Furthermore, the described change will be characterised with regard to types of change that occurred – in HI terms – in order to provide a description of SBT's rise.

6.1 Incremental Change: Legal Versus public

In order to triangulate the legal and public arena, it is first necessary to consider the types of data that legal papers represent. It is evident that legal papers are not as rich in material as journal articles are. This is partly because legal texts are often vague and uniform, a quality that is necessary for their legislative purpose. Furthermore, medical education's transformation, if it can be discerned with any certainty at all, is not explicitly articulated. In contrast, journal articles are rich in argumentative force in favour of either one version of the medical educational reform or another, but are oftentimes unclear in their legal or workplace feasibility (see Chapter 5.4).

The triangulation of data from these two sources is, therefore, useful in providing a general timeline of the changes mentioned in both types of data, as well as in identifying the key legislative or other significant changes that are then discussed in the public arena to reflect the structuring that each professional arena provides for the other and, thus, the claims being put forward. The Federal Ministry of Health issuing the examination order, and the IMPP issuing the catalogue of questions for the written examinations, are central for the legislative change when considering the legal and public arena for medical education, while experts and expert associations and their output (i.e., in journals, issue public statements (see Chapter 3.1-2.)).

One approach is to consider critical junctures for the given field when considering a timeline of transformation. Two critical junctures can be considered to be important for medical education in the legal and public arena, one that has an internal origin and one that has an external origin. The earlier critical juncture, which had an endogenous nature, is the publication of the NKLM. The later critical juncture, which had an exogenous nature, is the COVID-19 pandemic and its changes to the educational system.

First, when considering the legal and public arena for medical education, a general development of education in Germany informs this relationship since the early 2000s: the shift towards competence-oriented education. This type of education gained momentum in the medical educational public arena in the late 2000s, resulting in a competitor-as-guideline to the governmental examination order: the NKLM put in place by the MFT and GMA in 2015 (Jünger, 2018). In this instance it was not the governmental structure, but the medical educators within their expert associations who organised towards this comprehensive document. It can function as a "catalogue of items" (Gegenstandskatalog) (i.e., a reference to the individual university institutions about relevant educational content).

The NKLM, which focused on the concept of competence, arguably had the (unintended) consequence of highlighting the existing regulations' inadequacy and can, therefore, be considered a critical juncture for the Federal Ministry of Health's reform inertia. This paper effectively undermined the prevailing legal standard, compelling a reaction from the legal arena because of its comprehensive nature and its contradiction of existing rationale. In this regard, the medical educational profession wielded significant influence, exerting substantial pressure on the legal arena from the public arena. While the MFT and GMA succeeded in establishing professional jurisdiction in the public arena, the Federal Ministry of Health responded by setting up an expert commission to (in their turn) produce a more competence-oriented document, the so-called "Masterplan Medicine 2020" in 2018 (Wissenschaftsrat, 2018), which emphasises competence-based learning even more and calls for a reduction of the NKLM's questions and topics by about 15 to 20% (Jünger, 2018).

The Masterplan Medicine 2020 proposes a shift towards the utilisation of the NKLM as a shared reference framework for future curriculum design and replaced the IMPP item catalogue. The document further calls for enhanced collaboration between the IMPP and academic institutions (Wissenschaftsrat, 2018, p. 10). In turn, the Masterplan Medicine 2020 is regarded as a significant undertaking, necessitating supplementary financial support for medical universities and study hospitals (Wissenschaftsrat, 2018, p. 14), which is indicated to be financed by the ministry. The Master Plan emphasises that the NKLM is a joint document of the MFT and the GMA, but mentions that other professional associations and institutions were also involved in consultations, such as the BÄK and the BMG itself (Wissenschaftsrat, 2018, p. 19).

It is central to recognise that this argument plays an important role in establishing the NKLM as an authoritative entity. This emphasis serves to circumvent any notion that the NKLM emerged as an unexpected development, but rather that it was a projection that was foreseeable by the ministry, as well as by other relevant actors within the field.

Finally, SBT is mentioned explicitly as part of the professionalisation of medical education – with the exact wording of a "professionalisation of teaching at medical faculties" (translation J.D.) – with OSCEs as examination formats, teachers receiving additional didactical training (i.e. a Master of Medical Education) and the establishment of institutions specifically for education, namely the simulation training centres and Skills Labs, which are regarded as central to contemporary medical education (Wissenschaftsrat, 2018, p. 22). Nevertheless, the Masterplan Medicine 2020 has been the subject of approval and criticism in public by the GMA's committees, particularly with regard to its feasibility and the necessary funding for the medical faculties' reform (2019-V36-I4 (11)). This notwithstanding, the Masterplan did not come into effect initially, due to the COVID-19 crisis, but the proposed reform in the form of a new examination regulation has been put on hold since then (and at time of writing) due to other issues on the Ministry's agenda (2025); this has given rise to much public debate.

Second, when searching for Simulation-Based Training within the two types of legislation considered, I found that the examination orders did not mention SBT itself, but instead mentioned the possibility of using simulation patients and simulators for practical state examinations in exceptional situations (AppO 2021, p.7). Exceptional situations are not defined further in this instance, just as there is no further reasoning relating to the inclusion of SBT's commodities; however, considering the public arena's debate during the COVID-19 pandemic, and if or how medical education could be continued under the new hygiene protocol, this shift within the examination order is very likely originating from the necessity to provide universities with possibilities to react to the crisis. SBT is, therefore, included as an alternative with the need for explanation, not compensation for the usual practical examinations. Furthermore, when considering the educational practice, given that it is also reported within the public arena by the educational experts, new procedures, especially using digital solutions became central to continuing teaching and testing with SBT (i.e., training communication skills by using video conferences with SPs) (2020-V37-I7 (Covid) (12), P. 3; 2022-V39-I4 (2), P. 3).

However, if the analysis were to remain at critical junctures, then several important aspects of the rise of SBT between the 1980s and the 2020s would remain undetected as a result of their invisibility. It is imperative to also consider incremental changes in order to achieve a comprehensive understanding of SBT's stabilisation in its implementation at all

German universities (Lohse, 2024, pp. 95–98; Stark, 2018). For this aim, not only are the state examinations at the federal level important, but so too are the study regulations and curricula for each university at the Länder level. As described previously, state examinations did not mention any SBT until very recently. However, a much earlier mention of SBT and Skills Labs, namely in 2003 compared to 2021, can be found within the study regulations. Nevertheless, a distinction must be made for the individual universities in the concrete implementation of SBT (i.e., in use for teaching as either mandatory or facultative). In some cases, students can freely choose to spend more time in Skills Labs compared to exams, where some universities mention the use of simulators or simulated patients for exams (see Chapter 5.3).

In contrast to the legal arena, the JME articles show a discussion of both teaching techniques and examination mode much earlier, from the very inception of the journal in 1984. An article in the very first issue of the JME in 1984 speaks of a simulation patients' programme in Münster, Germany, meaning the usage of actors and actresses to help students simulate communication with patients or examination of bodies, which is a major component of modern Skills Lab training (1984-H1 (3)). Furthermore, there is a very early example of a description of a room decorated to look like a doctor's office to practice these situations with simulation patients in Mainz in 1993 (1993-H1 (9)). Today, that could be very likely be one of the rooms found in a modern Skills Lab. There are also several accounts for building and using materials and objects, to tinker with them on a medical models or dummies, such as the usage of a dummy equipped with a speaker playing heart sounds that students can listen and learn more about using stethoscopes (1986-H1 (2), S. 2; 1994-H1 (1), P. 1). Even machinery used to train better hand-eye-coordination or self-made models of body parts can be found within the journal papers from the early 2000s (2007-V23-I1). All of these examples speak to the tinkering and slow professionalisation of Simulation-Based Training, mainly from exceptionally motivated or interested staff, and the given institutional willingness to try out different teaching, and later examination, styles as Layering of existing structures, while not abandoning traditional teaching methods.

Examples are given within the public arena to advocate for SBT training and include the feasible use of commodities. They put pressure onto the legal arena, both directly and indirectly, while simultaneously referring to it, commonly by mentioning the examination order. The narration is that the 2002 examination order did ultimately offer opportunities for curriculum development (i.e., model study courses and to reorient towards practical training). However, the lack of a definition for practical skills, and moreover how they should be implemented into curricula, is oftentimes criticised. This came not only from the NKLM, but was also featured in earlier papers such as the GMA Practical Skills Committee's Consensus Statement and are an expression of the desire to enhance the educators' jurisdiction and standard-setting processes. The Consensus Statement is a milestone that is incremental, similar in structure to the NKLM, but that did not yet have the same impact on the educational community. It is the manifestation of competence as the core of educational development and it follows on from the NKLM to an interim end with the Masterplan Medicine 2020, should have been the foundation for a new examination order in 2019 after the smaller changes of the examination order of 2012 and that has not been resolved at time of writing (2025/2026).

It is noteworthy that the involvement of expert groups in the development of reform proposals by the Ministry is not a novelty, but instead occurs regularly, in 1993, for example, when a report on the activities of the expert groups was published that once again demonstrated the strong commitment to the medical profession's self-organisation and the necessary input being provided for the complex German health and education systems (1993-H2 (9), p. 1).

Can a development of three phases be described to ground these developments on the national level back to the institutions of SBT?. First, the beginning and pioneering stage lasting from the 1980s to 1998. Here, SBT's functions are beginning to be discussed and the call for Skills Labs starts in earnest, but only provisional simulation-rooms appear (1984-H1 (3), p. 1; 1986-H1 (2), p. 4; 1993-H1 (9), p.1; 1994-H1 (1), p.1). At the same time, educators also advocate for greater freedom within curricula to implement model study courses and to critique the BMG for its reform inertia in medical education (1996-H2 (2), p.1-2). The second phase then starts with the first Skills Lab in Germany, in Berlin in 1999, and lasts until 2011. Within this timeframe, the establishment phase of Skills Labs, more and more Skills Labs are being founded and the debate on competency-based education takes centre-stage for educators. In 2012 the NKLM is published and the consolidation phase of Skills Labs starts and lasts all the way until the end of the analysed material in 2022. While the overall restructuring of medical education is only just beginning, many of the Skills Labs and their SBT programmes have had several years to stabilise and are now concerned with quality standards (2016-V33-I4 (10), p. 2; 2016-V33-I4 (10), p.1; 2018-V35-I3 (11), p.3; 2019-V36-I3 (5), p. 1; 2019-V36-I3 (4), p. 1).

However, the professionalisation of medical education, towards practical skills, competence and SBT as methodology for the aforementioned goals, is not yet completed within the three jurisdictional arenas (Abbott, 1988; 1991, p. 20). The missing new examination order is evidence of this absence. However, the transformation described did show unintended consequences that rule makers produce and face because the individual university circumstances are more complex than the standard-setting of educational experts or governmental structures can anticipate within their regulations (Streeck & Thelen, 2005, pp. 13–15). Accordingly, the current organisational formats of Skills Labs varies across Germany (see below: legal versus workplace arena).

As a reminder, in the first phase of pioneering, portrayed in the public arena, no reports of SBT or Skills Labs could be detected in the legal arena. In the second phase of establishment, singular universities mentioned SBT and Skills Labs within their study curricula, but still no report could be found within the state examination. Lastly, at the very end of the third phase of consolidation, the examination order of 2021 mentions SBT commodities for the first time. SBT's progress SBT as reported in the public arena is, therefore, ten to twenty years further advanced than it is in the legal depiction.

To conclude the triangulation of the public and legal arenas, it can be observed that the theoretical input by Abbott's theory that the public arena is structured in its discourse by the legal arena holds strong analytical evidence for the professionalisation of SBT, as well as for the public arena that puts pressure on the legal arena through publication activities and the experts' quasi-legislation, like the NKLM, whereby medical education as a subordinate profession to the medical profession improves its status overall: "One competing group raises its standards or organizational efficiency and thereby threatens the public and perhaps legal jurisdiction of its environing competitors" (Abbott, 1988, p. 97).

6.2 Incremental Change: Legal Versus workplace

The following analysis of the Skills Labs' institutional composition is intended to provide an insight into the ways in which the legal arena is reflected or re-issued in the context of the workplace. The federal state is a key factor in comparing the legal and formal embeddedness of the universities or medical faculties of interest. However, there is also considerable diversity within this group of the eight Skills Labs. This concerns the faculty's composition in relation to the study programme, which can either be a regular or a model study programme (see Chapter 3.2. and Chapter 5.1.).

It is also important to determine how Skills Labs are described in legal documents. This includes how they are described in terms of teaching content or subjects, assessment formats and general information regarding the SBT and its commodities. This distinction between teaching and testing is relevant because the examination order provides only an outline for all German medical faculties, each of which must decide on how to implement it in their own programmes and, possibly, to further define (educational) rules, to use HI terms. The examination order must, therefore, be broad enough to be applied by all institutions, but specific enough to avoid legal challenges by students or examiners.

The study regulations, conversely, represent individual institutions with different histories, sizes, specialisations, collaborations, etc., with as much interest in well-functioning examinations as in well-functioning teaching. Furthermore, the workplace arena becomes important for these functions, given that educators within the Skills Labs must take the given rules and implement them, while possibly holding divergent stances, as voiced within the expert interviews.

Chapter 5.1's analysis of the legal arena provided an overview of the diverse transformation of medical departments in North Rhine-Westphalia. For instance, Bielefeld's medical faculty was only established recently, in 2021, whereas locations such as Aachen or Cologne have been offering model study programmes with innovative educational outlines, including SBT and Skills Labs, since the early 2000s. Nevertheless, if the formal outline of curricula is considered in isolation, then it could be argued that most locations are merely mid-range adopters of the second consolidation phase of Skills Labs after 2010. Such descriptions, however, are misleading due to the legislative documentation's inherent inertia.

Following the interdependence of the legal and workplace arenas, and according to Abbott, the legal arena structures the working conditions; in HI terms, it provides the rules. In turn, the workplace arena can also undermine formal rules in their everyday practices; in HI terms, reinterpreting or bending rules(see Chapter 2.1). Two aspects are central to triangulating the legal and workplace arenas. First, how does the transformation of study programmes, and within them the SBT structures, implement the teaching and testing procedures? (i.e., when and how do interviewees report the influence of legislation?). Second, how did possibly rule bending or undermining behaviour result in medical education's transformation towards SBT, which is not reported within legislation?

Firstly, the design of study programmes' impact on teaching content, particularly in relation to SBT, becomes apparent when analysing the educators' interviews in terms of both teaching and testing. As with all legal documentation, these documents only provide limited information about the reasoning behind the legal changes (i.e., the decision-making process of a medical faculty to transition from the regular study programme to the model study programme). In the interviews, the educators provide insight into the advantages and disadvantages of both curricular versions. However, in order to maintain the anonymity of the interviewees, I will refrain from providing detailed descriptions of each location's circumstances. This decision is informed by the limited number of individuals associated with SBT and the number of locations analysed in depth, so that individuals would not be identified. Instead, the focus will be on presenting the arguments and, potentially, providing a general account of the transformative process.

The distinction between the model and regular study programmes is generally characterised by the separation of the pre-clinical section of medical studies, which is mainly concerned with the 'big subjects' like anatomy, biology, chemistry and physics and the clinical section, which includes more practice-oriented courses. As per SkL05, this can mean that students do not visit the Skills Lab for extensive mandatory courses until later in their studies. Simple models are used in the pre-clinical phase for some basic skills, like sonography and anatomy. However, most of the Skills Lab's use occurs after the fifth semester, during the clinical phase, when students simulate patient contact with SPs alongside bed-side teaching (Transcript SkL05, Pos. 43–44).

Furthermore, with regard to legislation, one interviewee remarked that the pressure and learning goal of practical education has risen steadily since the early 2000s, resulting in some universities deciding to offer model study programmes and, therefore, consequently combining preclinical and clinical content, whereas other universities reformed their curricula within the regular study programmes (Transcript SkL06, Pos. 17, translation J.D.). This pressure towards reform resulted in SBT's introduction, according to SkL06 (ibid). Another interviewee echoes this observation of change, describing how many locations changed their curricula from regular to model study programmes after the demand for competence-oriented education, particularly when communicative competence increased. They link this development to the NKLM and argue that this was the rationale for medical faculties to start teaching projects, which stabilised into Skills Labs and its subdivisions for communication, which are nowadays their SP programmes (Transcript SkL09.1+09.2, pos. 19–20). This reiteration by several interviewees marks the late 2000s as a significant turning point in medical education, as evidenced inter alia by the NKLM's establishment, in connection to the reorganisation of curricula at various faculties towards competence-orientation (Transcript SkL03, Pos. 13, 45–46).

In a similar vein, there seems to be an emerging pressure to restructure curricula to be more practice- and competence-oriented with the aim of preparing for future regulations, such as a new examination order. This outlook necessitates either a proactive reform of curricula within the established framework of a regular study programme (Transcript SkL12, Pos. 26–31), or the implementation of a model study course, with the anticipation that fewer changes would be necessary after the fact (Transcript SkL08, Pos. 9). This suggests that the strategic thinking of the actors involved is not divergent with regard to the transformation of education, particularly in relation to working within the Skills Labs. They have been designated and adopted the role as forerunners and are supporters of innovative didactics.

When asked about possible future distinctions between the regular programme and the model programme, in the event that the examination order places a strong emphasis on competence orientation, SkL12 answered that the programmes would most likely be similar (Transcript SkL12, Pos. 31–33). However, both strategies also speak to a reactive or proactive nature and towards 'bigger politics' (i.e., the macro- level of legislation of the examination order), without emphasising the design of such legislation in their own interest. It is noteworthy that other interviewees self-identified as 'not political' and expressed a lack of engagement with the discussions surrounding the curriculum format beyond their faculty. These interviewees indicated that these discussions were not pertinent to their day-to-day work and, therefore, were not of interest to them, framing them as, for example, "That is a political matter [...] That is on a level, where I do not participate in, I only hear about the results" (Transcript SkL12, pos. 27–29, translation J.D.) or where see themselves as not fit (Transcript SkL09.1+09.2, pos. 114) or not welcome to participate (Transcript SkL03, pos. 33).

Nonetheless, these remarks do show that there are other active participants in both agenda-setting towards polity and policy, through taking stances and being active in the public arena. This speaks to the possibility of border-crossings between workplace and public arena, while it is framed as similarly legitimate to not participate in the bigger discourse.

In conclusion, the composition of study programmes is central to each individual faculty, even where there is no interest in developments beyond the home university. Moreover, the restructuring process is often associated with costs, either in the form of institutions such as the Skills Lab with their personnel and equipment, but mostly through the time investment of producing learning goals and curricula and, ideally, the didactical briefing of teachers. The curricula must be aligned in several respects: Firstly, with regard to the adjustment to the examination order, as a macro-level of the legislature for the whole of Germany, and secondly at the federal state level, whereby each faculty has a self-governing right, whether to implement a model study programme or a regular study programme for example. Thirdly, however, the curricula must align the learning objective with the form of teaching and assessment, which is also known as constructive alignment. Constructive alignment is a concept frequently referenced by educators in relation to SBT (Transcript SkL06, Pos. 4–5; Transcript SkL09.1+09.2, Pos. 24; Transcript SkL15, Pos. 25). The interviewees do, when reflecting on the formal circumstances the Skills Labs have evolved in, oftentimes refer to the lack of uniformity, while emphasising that the implementation of a Skills Lab is universal nowadays (Transcript SkL01, pos. 13, translation J.D.).

The interviewee subsequently provides a detailed description of the organisational structure of certain universities, which includes different study hospitals that are connected to one faculty (ibid). This results in the presence of multiple 'skills rooms' at different locations, rather than a single, comprehensive Skills Lab. How some Skills Labs only offer mandatory training, while others have open spaces for students to train on their own on a facultative basis, is also described. Others are organised exclusively by student tutors. It is evident that these varying examples are all related to the given curricula and to the interpretation of the examination order (Transcript SkL01, Pos. 94–95).

This variance holds true for the Skills Labs in North Rhine-Westphalia (see Chapter 5), while the interviewee also provides examples from across Germany. They attribute the variance to the historical status of the university in question, i(.e., the extent to which education is funded or valued at the respective location) while emphasising that no version is superior to another: "Ultimately, there is no systematic approach, if I may say so. So, in the end, you can't say Skills Labs are better managed if they are one way or the other" (Transcript SkL01, Pos. 28, translation J.D.). In response to the question of how new the kind of practical teaching now taking place in the Skills Lab is for medical education as a whole, another interviewee emphasised that the Skills Lab itself offered an integration of practical teaching efforts, concentrating efforts in one institution rather than in individual course offerings. Historically, this underlines the pioneering work of individual educators, but it also testifies to the professionalisation of medical education overall (Transcript SkL02, Pos. 104–105, translation J.D.).

The advantage of Skills Labs is that they are formalised locations for practice-oriented teaching that commits to the acceptance of errors without endangering patients. This approach has reduced stress among students and facilitates constructive feedback within a single institution since, in this case, the mid-2000s (Transcript SkL06, Pos. 17, translation J.D.). As outlined in Chapter 3, teaching and testing hold different status within the legislation, due to the need for a narrower definition of assessment within the curricula. Teaching can allow for greater variance, while testing needs to be unified in order to ensure fair circumstances for every student. Accordingly, the standardised procedure of the OSCEs is an effective means of integrating SBT into the curriculum and suggests that SBT procedures are also used in teaching to prepare for the assessments. Nevertheless, the formal organisation of assessments between different locations is not as uniform as educators suspect; this is similar to the predicament found within teaching described previously.

SkL09.1. describes that staff at their Skills Lab occupy an intermediate position between the university hospital and the university by being employed via the dean's office of the medical faculty, specifically for the OSCE exams, but this construct is not universal at all universities. This variance is emphasised as surprising by SkL09.1 because the formal framework of the examination order, or at least the MFT as the union of all medical faculties, could suggest a uniform way forward for OSCE assessments (Transcript SkL09.1+09.2, Pos. 80, translation J.D.).

The interviewee's astonishment is noteworthy in the context of the relationship between the legal and the workplace arenas. This is because there is no uniform assessment framework, even though different concepts both exist and are discussed. This aspect of inconsistency is represented in the difference between the all-encompassing examination order and its examinations and the examinations conducted by the individual faculties according to their curricula. Furthermore, the variance in institutional composition can go even further when considering one case in North Rhine-Westphalia where the teaching and testing proceedings are divided into two respective Skills Labs, which have individual directors and agendas. This raises a fundamental question about the definition of Skills Labs as training facilities about whether assessment is the sole activity taking place. Furthermore, it raises the question of how educators view rule-making and rule-taking in the field of SBT.

The second focal point of the triangulation of the legal and public arenas considers the rule-making and rule-taking aspect that HI attributes to institutional frameworks. In terms of professional jurisdiction, it examines how legislation is perceived and enacted by educators. This pertains primarily to the examination order and to the NKLM as overarching frameworks in the context of medical education and within the specified timeframe.

First, the reorganisation of practical training of medical education came alongside the competence-orientation, but furthermore with the political rationale of having graduates become a part of the workforce sooner. Interviewee SkL06 exemplifies by discussing the removal of the 'doctor on internship', which did not give graduates the status of full doctor but trainee, who would still be supervised and, therefore, had a longer transition time from studies to full professionality (Transcript SkL06, Pos. 17, see also Chapter 3.2.). This internship status of 18 months after graduation was taken out of the examination order in 2004, and instead the last year of studies involve small, rotating internships within a 'practical year', after which time the graduates gain full assistant doctor status after their second state examination (Penter & Augurzky, 2014, p. 79). This example, which is not directly connected to SBT, showcases the contradictory consequences of rule-making that was meant to relieve pressure on the health care system, due to the lack of personnel, while forcing medical education to practical training elsewhere.

Second, SBT is the chosen path in this instance, while offering a longer didactical debate since at least the 1980s, as shown in the journal articles. Moreover, what the turn towards SBT showcases is the self-organisation of medical education towards standards of practical skills and training, culminating in the NKLM. As SkL01 remarks, it seemed inevitable, when looking back on the development, that competence would be the central aspect that education had to focus on: "It's not important that we want to teach them something. What matters is what they can do afterwards. And if we focus on this idea together with the issue of competency-based learning, then sooner or later there will be no getting around the fact that we need to systematically develop competencies which also include skills." (Transcript SkL01, Pos. 88, translation J.D.).

However, as mentioned in Chapter 3.2., competencies and skills need definition and concepts, how to be implemented in curricula, and furthermore how to be assessed, given that they can include latent abilities. While the NKLM offers a framework for all medical faculties, each one can (and will) choose focal points based on their individual tradition, formulating their own learning catalogues based on the NKLM or using other competence-based concepts, such as the EPAs, as a basic concept to introduce in their case descriptions of SBT, as SkL15 describes for their location (Transcript SkL15, Pos. 21). This means, that the NKLM's rule-making was a self-organised effort that, like with any formal framework, did find reworkings and different interpretation by each university's rule-takers.

This process, as reported by the interviewees, is also described in the literature, which considers how SBT should or could be implemented in university curricula, ultimately describing educational goals in terms of competence theories, which speaks to the close intertwining of these two fields (Ahlers, 2018). While there are theoretical didactic concepts of competence, educational professionals also need to define competence for themselves in order to organise their institutions and frameworks, the Skills Labs and SBTs, which often leads to atomisation into smaller steps that educators want students to be proficient in.

The gap between workplace reality and legal frameworks, especially regarding the examination order, can nevertheless be considered to be bigger than indicated by theory. Referring to the HI, the re-interpretation of rules by rule-takers is a given; however, in the instance of SBT, the institutionalisation is far-reaching in comparison to the non-systematic approach towards SBT by the legal arena. Interviewees refer to the shift towards practical teaching and competence-based assessment in their curricula, by proclaiming that if students cannot act professionally after the study programmes, then they cannot be considered to be competent (Transcript SkL01, Pos. 44, translation J.D.). This is a definition that is a near-identical replication of a definition frequently employed within the medical educational theory, namely that of Miller's pyramid (Miller, 1990).

As a result of this rationale bound to the academic knowledge of didactics, the medical educators inevitably both hold onto and emphasise their autonomy as university educators repeatedly, pushing against unwanted changes within the examination order or the exam composition by the IMPP by formulating position papers and unionising within associations, such as the GMA (Transcript SkL03, Pos. 33). This initiative's agenda-setting nature has consequently resulted in a shift in the Federal Ministry of Health's approach towards medical education reform, as evidenced by the commissioning of an expert committee to work on a reform plan, which resulted in the Masterplan Medicine 2020; this Masterplan explicitly included SBT for the same justification as described by SkL01 above (Wissenschaftsrat, 2018, p. 61).

Translating these considerations back into HI terms, the assessment of practical training in medical education can be considered a Displacement. However, this may not be applicable to the domain of teaching transformation because SBT can be regarded as a distinct format among others, as outlined previously. *This suggests a Layering of teaching methodologies in medical education, with SBT representing a particularly intensive form of Layering. Additionally, the Displacement occurring in assessment is not yet fully developed because the Masterplan Medicine 2020 has not yet been ratified in a new examination order as of 2025.* Nevertheless, educators are preparing and taking stances about the future examination order, and the GMA provides comment drafts that are made public. One such aspect, which would serve to consolidate the Displacement of the majority of other practical assessments, includes the incorporation of simulated patients within the state examination's practical exams. This approach would align with the methodology employed by certain faculties in their own faculty intern's OSCE assessments.

When interviewees are asked whether they believed the changes to the future examination order were in favour of SBT, and therefore that they offer fundamental reform of medical education, they were reluctant because of the funding of such a shift. SkL14 emphasises that SBT's financial implications are not limited to the initial costs of setting up the necessary facilities and purchasing simulators, but also extend into the ongoing costs of maintaining these commodities, employing SPs and staffing the SBT trainers themselves. These are costs that are often underestimated by medical faculties and that would be correspondingly high in the event of reform (Transcript SkL14, Pos. 98, translation J.D.). This is due to the fact that the examination order would suffice for medical faculties to also request a higher level of financial support from the government in order to expand the Skills Labs to an extent that would enable them to coordinate practical state examinations for each student.

Another interviewee, SkL09.1, has similarly expressed scepticism regarding the prospect of further reform to medical education in the direction of SBT. This is due to the fact that the programme is costly and, as is the case in other fields, the prioritisation of reform in education could be diminished in favour of other issues (Transcript SkL09.1+09.2, Pos. 114).

Other interviewees are less pessimistic and expect the NKLM to play a central role in the examination order. Even with extensive changes, though, such as reducing the number of competencies formulated and limiting specialist medicine foci: "The NKLM will most certainly be shrunk again, because this very ambitious view of every little corner of medicine, which is also driven by specialist societies, cannot be maintained if it is taken seriously. On the other hand, the competencies that I have just mentioned, starting with hands-on skills and so on, are part of the NKLM and will not be completely eliminated." (Transcript SkL01, Pos. 92–93, translation J.D.).

To "fight for such turf" (Abbott, 1988, p. 102), necessitates a meticulous evaluation of the training objectives that are to be incorporated and those that are to be excluded. As SkL12 states, if the learning objective no longer exists, then there is no test objective and, thus, the practical skills and Skills Labs can be eradicated as institutions (Transcript SkL15, Pos. 210). However, SkL08 also argues about the extent to which they can be justified, even beyond the individual students' abilities, in order to protect patients while reform efforts necessitate investment on multiple levels, including location, equipment, educational, personal and administrative manpower to formulate curricula on the federal and Länder levels. They observe the tense working conditions within the current health care system, with a significant proportion of graduates deciding against pursuing careers as medical professionals. This issue could not be addressed by increasing the number of students, but instead necessitates the implementation of appropriate training and working conditions overall (Transcript SkL08, Pos. 9).

6.3 Incremental Change: Public Versus workplace

This final subsection will triangulate the public and workplace arenas again in relation to SBT's implementation in the eight case study sites and in the teaching and assessment formats used. Interviewees frequently refer to the public arena and to actors, such as the GMA and its position papers, because they themselves cross the public/workplace border, as was alluded to in the previous section. However, a clear-cut outline of SBT becomes necessary towards the goal of jurisdictional claims in the public arena in comparison to the workplace arena, where interviewees can formulate contradictory developments and influences on their everyday work. As Abbott states: "The tremendous inconsistency between the public and workplace realities of professional life has grave implications, not only for professional life in the workplace, but also for the public jurisdictional structure. The diversity of professional workplaces must be reduced, in public, to a clear, simple picture of straightforward jurisdictional relations between undifferentiated wholes" (Abbott, 1988, p. 84).

Consequently, cases of interest include those in which the public representation of SBT contains such inconsistencies. In the following case, the focus will be placed on SBT's acclaimed rise because it is suggested that this field is less unified or professionalised than has been claimed. In this particular instance, however, the fundamental cause of the revision is the definition of the training in question and because of the institutions responsible for its delivery. In 2016, a special issue on clinical skills was published, in which the editorial made the following claim:

> "One particularly successful major project in this area is documented in the present special issue 'Clinical skills': 20 years of development of 'skills labs' in the D-A-CH region (Germany/Austria/Switzerland)! Clinical practical skills have been taught before, according to training regulations and laws. However, the fact that an impressive number of laboratories [...] have been set up for the systematic and longitudinal training of skills can only be interpreted as part of a transformation process, i. e. the professionalisation of medical education. 1996 marked the beginning of the establishment of learning centers for the teaching of patient-centered skills" (2016-V33-I4 (14), P. 1).

Whilst the claim is broadly accurate in its conceptualisation and analysis of the transformation, it does so with reference to specific locations in which Skills Labs for the training of clinical skills have been established. The claim names 1996 in Vienna, followed by 1999 in Berlin, Heidelberg and Frankfurt/Main in 2000, Cologne, Mainz and others as all occurring in 2003 and beyond (ibid). However, an examination of the dates of the first implementation of a Skills Lab or similar institution for Simulation-Based Training in Switzerland, Austria or Germany is contradictory. A study by Fandler from 2012 (see figure 9) provides a notable example, as it requested information about the founding years of all Skills Labs in the three countries mentioned previously. The year 1992 is mentioned first, presumably referring to Vienna's Skills Lab.

Figure 9: Simulation centres in Austria, Germany and Switzerland (Fandler, 2012, p.23)

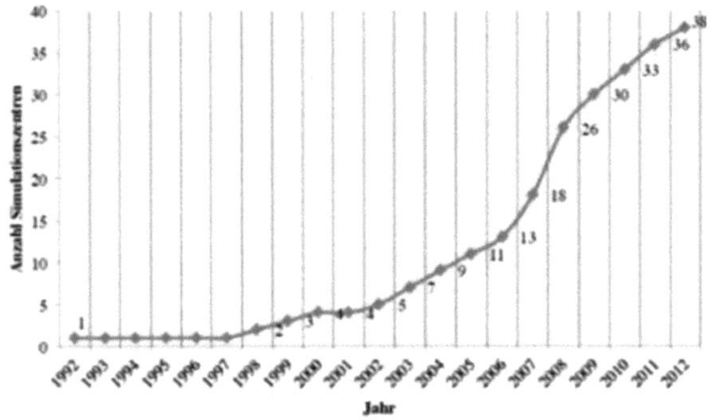

This finding stands in contradistinction to the JME's anniversary of 1996. Moreover, this finding aligns with another report that was published in response to the JME's initial article, which also corrected the implementation date for Vienna as follows:

> "Since 1989 the study site in Vienna's Medical faculty has started its operations step by step inside the New General Hospital. In the editorial on 'training of practical skills' the operational start of the 'Skills-Lab' is only recognized in 1996. We have therefore checked the developmental history by reference to distinctly diverse documents: design papers, project reports by the study site and university calendars, all of which only exist in print form for this era and are mainly "grey literature". [...] According to those sources it needs to be recorded that an elective course on training of practical skills was first announced for the winter term 1992/93 [...] This investigation has revived the memory that systematic skills training has started in Vienna almost 25 years ago and has brought back a number of questions: What were the supporting

> factors that had contributed to this early innovation? How has this institution developed from its role as an "early adopter"? What is the current status of scientific documents and publications that are difficult to access? How significant are statements concerning the institution's seniority anyway? After all various elements for the training of practical skills have been used for a long time, such as obstetrical phantoms, support devices to train auscultation or ophthalmological models (e.g. in Berne, Switzerland). In this respect one would rather have to pose the question why it has taken so long to transform those early approaches into "major projects" – resp. which factors have induced the vigorous development in the "D-A-CH Region" (Germany/Austria/Switzer-land) – approximately since the turn of the millennium" (2016-V33-I5 (1), P. 1).

The above feedback seems to indicate that the Vienna Skills Lab was implemented in 1992, not 1996, but it also shows that celebrating anniversaries can become an agenda-setting event in the public arena, underlining SBT's importance. In this case, the response indicates that SBT had already existed prior to the anniversary date provided. Moreover, this answer and the various questions it raises show that the image of SBT, and the Skills Labs as its institutionalisation, is not as uniform as either the editorial article or the quote from Abbott might suggest.

Furthermore, it is important to note that the quantification of the implementation of Skills Labs, as discussed in the study by Fandler and in another report by Segarra et al. in the JME (2008-V25-I1 (25)), is not straightforward. However, it can be demonstrated that the number of Skills Labs does increase steadily from the 2000s onwards, with a further steep increase from the 2010s onwards; this is in line with the phases identified previously in this study. The present study is focused on Germany and, therefore, the initial founding in Vienna is of lesser importance. However, the article in question and the interviews conducted provide the view that the first German Skills Lab was founded in 1999 in Berlin (Transcript SkL04, Pos. 3; Transcript SkL01, Pos. 86), followed by several Skills Labs being founded in the years that followed.

In conclusion, the 2016 article provides both a narrative and a historical description. The authors establish a direct line and a steady growth trajectory towards the implementation of practical, skills-oriented training, for both SBT and Skills Labs, and outlines the self-organisation of educators towards the NKLM (2016-V33-I4 (14), P. 2). While the narrative is broadly applicable, it does reveal the authors' agenda with regard to their social claims. Further investigation is required to better understand how incremental changes were characterised by parallel developments and a back-and-forth process, focused in this case on North Rhine-Westphalia. The following sections aim to demonstrate how North Rhine-Westphalian locations discuss their SBT within the context of JME articles disseminated within the public arena. This will be followed by a triangulation with statements from educators in the workplace arena and concerning both SBT's introduction and rise.

The following section will list the content of articles that directly discuss SBT at the North Rhine-Westphalian faculties, provided that they have not been mentioned previous and where they have relevant output. The articles listed below refer exclusively to projects by or about the SBT programmes and Skills Labs; however, not all locations could be represented. It should, therefore, be noted that the list is not exhaustive. Authors from the institutions mentioned may also be co-authors of other articles in other functions, as a result of their chairing a GMA committee for example, which is not the focus here.

- The Aachen Skills Lab is considered to be among the most established of its kind. It is referred to by the acronym AIXTRA, an abbreviation for Aachener interdisziplinäres Trainingszentrum für medizinische Ausbildung. The Skills Lab's establishment is documented in an article from 2010, which details how the initial concept and financial resources were obtained in 2005. At that time, individual SBT courses could be conducted in shared rooms, indicating that an independent institutional framework had not yet been established. However, SBT had already been initiated. This was followed by a planning phase and the move into their own rooms in 2007 (2010-V27-I1 (6), p.6). In addition, as was revealed during both the interviews and location visit, there is now a completely new and even larger building for the Skills Lab alone. Aachen has no particular strong suit when it comes to SP or simulators; both are used frequently and visibly within SBT and, accordingly, during the pandemic SP training was not abandoned, but rather continued in an online format. Students gave this digital teaching with SPs an overall very good or good rating (2020-V37-I7 (Covid) (10), p. 1).

- Bonn is also mentioned in a deep dive into SBT's very early beginnings in North Rhine-Westphalia, but this location's importance should be mentioned here. Bonn is mentioned as using SPs (1984-H1 (3), p.1) and self-constructed simulators (1986-H1 (2), p.2) through a cooperation with the University of Münster as early as the early 1980s. However, these are individual educators establishing SBT, not complete programmes or Skills Labs. The literature dates the introduction of peer-to-peer courses with SBT back to 2012 and it was used to prepare students for their oral exams. This programme, linked to the Skills Lab, provided preclinical students with the opportunity to practice in preparation for the state examination (2018-V35-I1 (4), pp. 2–3). Moreover, Bonn is cited as one of two North Rhine-Westphalian faculties that are referenced in an overview article about the competence-orientation of medical education and about the subsequent restructuring of curricula. In this instance, the focus is on communicational skills, which the examination order has been requesting since the early 2000s. It is evident that various faculties employ learning tools such as role play, video analysis and SPs in both facultative and mandatory courses (2011-V28-I4 (4), S. 2).
- Bochum is mentioned as the second faculty, located in North Rhine-Westphalia, which also uses all of the SBT methods mentioned previously (ibid).
- Düsseldorf is particularly strong in the use of SPs to both teach and assess communicative competence, as evidenced by several reports in the JME articles and interviews. One such study involved the utilisation of SPs and actual patients for a formative evaluation of communicative abilities (2016-V33-I5 (2), P. 6). Moreover, the SP

programme's initial implementation, as a component of the Skills Lab was documented in 2012, including OSCE assessments, which was considered successful but complex in terms of training not only the SPs towards SBT, but also the clinical teachers themselves (2012-V29-I1 (6), p. 5).

- Essen is another site in which SBT with simulators and SPs is well established. In a manner similar to Bonn, digital solutions for SBT's continuation had to be found during the period of the COVID-19 pandemic. In Essen, this resulted in the conducting of online OSCEs with SPs via videoconferences (2022-V39-I4, pp. 6–7).
- As was the case in Aachen, Köln/Cologne was likewise an early adopter of SBT and, consequently, established their Skills Lab (called KISS, short for Kölner Interprofessionelles Skills Lab und Simulationszentrum). The Skills Lab, which has undergone both relocation and expansion since its inception in the early 2000s, has also established specific SBT standards and has exerted influence within the public arena. One example is the communicative skills training, again often the early reference point for SBT's introduction after the 2002 examination regulations, for which Cologne also uses SPs and developed an evaluation sheet to assess the training and students' progress as early as in 2005 (2005-V22-I1 (7), p. 2). Moreover, there are instances of SBT's content being reflected and of lectures on specific topics being re-addressed as a result (e.g., on pharmacology and the difficulty of communicating drug treatments to patients (2019-V36-I2 (2), p. 5)). Nevertheless, on-site visits also reveal the extensive use of simulators and other commodities.

In conclusion, it can be attested that the variance of SBT's commodities is consistent across all sites, with certain Skills Labs offering areas of focus and different kinds of mandatory and facultative training. In this regard, it may appear self-evident that simulated patients are more closely associated with communicative skills, due to their responsiveness and training in providing feedback. This association is similarly identified in the JME articles. In a similar vein, it may appear self-evident that simulators are more closely associated with hands-on skills, given their tendency to involve mechanistic processes that necessitate repetition and precision in manual dexterity. This notion is also reinforced by the articles published in the JME. However, observed instances of SBT frequently demonstrate a crossover between these seemingly disparate skills by requesting that students communicate with simple models or combine SP with add-ons for hybrid training. The distinction between communicative and hands-on skills is not entirely clear-cut and the attribution of one skill to one commodity is likewise unclear. It is only through the triangulation of different data points that this nuance can be revealed.

As stated in Chapter 5.2., the JME's initial issue already included both a report on and advocacy for the use of simulated patients and was written by Prof. Dietrich Habeck from Münster (1984-H1 (3)). In this report, Prof. Habeck calls for the consideration of alternative methods for the instruction of practical skills in light of the challenges posed by overcrowded German medical faculties. The utilisation of simulated patients is deemed a well-established international practice, and the Netherlands is cited as a prominent exemplar in this regard. Additionally, pilot projects at the universities of Bonn and Münster are noted for their contributions to this field.

Additionally, Habeck alludes to alternative SBT formats, namely the utilisation of simulators, which he designates as 'phantoms'. These can provide a high degree of fidelity and the opportunity to repeat a learning situation as often as necessary, akin to SPs, provided that they are introduced into medical education correctly (1984-H1 (3), p. 5). In addition to Bonn and Münster being mentioned within the context of our North Rhine-Westphalian sample, and thus being of interest as pioneers, the Thorax-Trainer as a simulator was constructed and used at the same faculties ten years later (1994-H1 (1), p. 1). Moreover, Habeck is of central importance to SBT's early development in (West) Germany (1994-H1 (1)). Habeck served as director of the GMA between 1981 and 1994, founded the JME in 1984 and was editor-in-chief until 1997 (2018-V35-I2 (4)). He was a professor and active educator in Münster until 1991. There are several speeches and reports printed that same year, 1991, about his retirement, the gap he will leave and the influence he had on medical education in Germany (exemplarily 1991-H1 (3)).

This influence of the reform of medical education towards competence, and therefore the introduction of SBT, is reflected in the "Münsteraner model" (1988-H1 (5), p. 4; 1998-H2 (7)). Several locations are quoted, between the 1980s and early 2000s, with model or reform projects, commonly being translated into model study programmes thereafter. However, Habeck was one of the central figures for the Münsteraner model and founded at his home institution, the Institute for Education and Studies (Institut für Ausbildung und Studienangelegenheiten), which is a subpart of the medical faculty in Münster today (Institut für Ausbildung und Studienangelegenheiten [IfAS], N.N.).

As reported in the JME, Habeck and his colleagues sought to emphasise medical education's didactic foundation and sought to initiate teaching projects that met international standards. Among these was the idea of using Skills Labs for practical, simulation-based teaching (1986-H1 (2), pp. 4, 14). However, it is important to acknowledge that the introduction of a reform effort and SBT's rise cannot be attributed to a single individual. Instead, a more comprehensive perspective is necessary, in line with methodological caution on historical analysis (Hines, 2011, p. 148). Nevertheless, it is important to acknowledge the role of individuals, such as Habeck, who proactively shaped the public agenda and, consequently, exerted pressure on legal and workplace standards to modernise.

Furthermore, following the initiation of a strategic approach, medical educators actively attempted to influence the political environment and reported on this in the JME in order to make it transparent that their efforts were not in vain by participating in networks and expert groups reporting to the Federal Ministry of Health (1993-H2 (7), p. 1). In conclusion, the JME as a journal was, from its inception, not only internationally oriented, but was also both deeply rooted and interested in learning theory and was as much a medical forum as it was an educational and political one.

For this reason, an overlap with the 'Murrhardt Circle' can be identified. This was a group of medical education reformers from various backgrounds who had no direct connection with one medical faculty and, therefore, claimed self-sufficiency. One of the founding fathers was Prof.

Habeck, who used the concept of competence very early on, and the communicative skills needed to become a doctor in specific, to argue for the reform of the MES. This was well-received by some, but not all, of his colleagues at the time, as the following quote on resistance to reform from 1995 shows:

> "The communicative competence has only played a subordinate role in the study programme reform to date. Exceptions to this are the analysis of the Medical Education Working Group (1989), which has already been mentioned several times, and the proposals of the German Society for Medical Education (GMA) in 1988" (Breucker 1995, p. 67 cited by Müller, 2012, p. 17).

Consequently, a significant proportion of the discourse in the 2000s and subsequent years was both influenced and inspired by the Murrhardter circle's contributions and other reform ideas concerning competence-oriented education, thereby saving the wheel from having to be reinvented (Stein, 2009). The examination order of 2002 and its subsequent modifications, which were influenced by these reform initiatives, have been implemented to a limited extent, thereby prompting educators to persist in modifying curricula and advocating for political change and leading to the diversification of facultative composition (Müller, 2012, p. 36).

What becomes clear from this deep dive into SBT's roots in North Rhine-Westphalia, and indeed as described in the JME articles throughout Germany, is the difference between, firstly, the advocacy of SBT and, secondly, its actual use in training and, thirdly, when SBT is then institutionalised (i.e., with its own sites as Skills Labs). This is an observation that is in line with Abbott's argument about the non-linear development of professions towards dominance, in this case with the rise of practice- and competency-based didactics using SBT as a training framework and becoming a central concern of medical faculties through the implementation, funding and ongoing support of Skills Labs, while at the same time grounding their claims in academic knowledge.

Interviewees frequently express a degree of uncertainty in relation to SBT's potential future. On the one hand, they emphasise SBT's utility and its commodities as learning concepts. SKL09.2 suggests that SBT has the potential to be significant beyond the context of medical education. For instance, in the field of teacher education, the training of teachers about how to converse with parents could be enhanced by SPs, thereby preventing an all-too-hasty entry into real-life experience. The interviewees also exercise caution with regard to the funding of SBT in the future (Transcript SkL09.1+09.2, Pos. 113–114).

At the opposite end of the spectrum, other interviewees adopt a more optimistic stance, unknowingly contributing to the argument of institutions' path dependency (see Chapter 2.2). SkL02 states that Skills Labs enjoy popularity among students in addition to possessing educational value. Therefore, comparisons between one faculty and the next, through rankings, there is a risk that any faculty would face a loss of students if they did not comply with having a Skills Lab. Ultimately, would the integration of SBT into state examinations necessitate a commitment from each faculty to allocate resources towards the development of such programmes, with a particular focus on addressing the prevailing challenges within the healthcare system? After all, the objective is to increase the number of students opting for a career in general practice or in other understaffed specialties (Transcript SkL02, Pos. 101).

The different expectations of Skills Labs and how they should be implemented into education become visible in several JME articles when opening the perspective towards all of Germany beyond North Rhine-Westphalia. The following report presents the findings of a survey conducted in the Skills Lab in Lübeck. The survey's primary objective was to address the discrepancies between the views of students and teachers with regard to the significance of particular skills and to determine the most effective applications of Skills Labs. It is particularly noteworthy that the survey was conducted both prior to and following the establishment of the Skills Lab, thereby indicating a meaningful shift in interest (2012-V29-I1 (7)). This demonstrates that, whilst there is a certain degree of path dependency in utilising the Skills Lab following its establishment, the precise form and content of the teaching cannot be guaranteed to be successful.

Similarly, another article provides different perspectives on the issue of teacher training as a quality standard, namely between educators and clinical teachers. The former are oriented towards learning theories and the correct application of learning methods, while the latter need concrete procedures for giving constructive feedback, thereby creating a positive learning atmosphere, understanding the needs of their students and understanding what makes a lesson successful (2002-H1 (14), p. 4). Once again, questions of constructive alignment between different status groups arise and raise questions beyond SBT and Skills Labs as complex institutions within a bigger framework. This underlines medical education as a rising, albeit subordinate, profession within medicine.

With regard to SBT's contemporary composition in Germany, other reports underline the variance found within the North Rhine-Westphalian sample provided. For example, a comparative study on the use of SPs in Germany, Austria and Switzerland states:

> "There are a variety of SP programs in German-speaking countries. While there are a few clear similarities (for example, feedback from SPs), many organizational and methodological aspects are handled Germany differently. Although this allows innovation and flexibility, it also weakens the didactic SP method in its standardization and thus in the comparability of quality. A certain degree of standardization and high methodical quality is of great importance, especially in scientific and faculty internal discussions and with a view to the use of SPs in high-stakes examinations which must be improved in the future" (2019-V36-I3 (4), p. 1).

Agreeing on common quality standards was identified as a key issue for the current phase in which SBT can be considered because while it is generally well established, it is not streamlined in medical faculties. A further survey of medical faculties' rationale for the implementation of Skills Labs also revealed that no pattern of argumentation emerges. No coherent arguments or unified reasoning could be identified by educators as to why Skills Labs are useful; however, this survey was focused on Skills Labs' risk management and economic efficiency (2018-V35-I3 (11), pp. 1- 3). The survey also attests to a lack of standardisation or certification of Skills Lab personnel, a concern that may be at odds with the educational goal of Skills Labs, but this similar to the analysis provided about the workplace arena.

6.4 Summary

The description of SBT's rise differs between the three professional arenas analysed with regard to the founding dates of Skills Labs, the use of commodities such as SPs and simulators, the didactic concepts, the assessment procedures (such as OSCEs and the study programmes, specifically model study programmes (see Appendix 8)). Accordingly, the triangulation provided in Chapter 6 offers a comprehensive perspective on SBT's rise and the associated professionalisation of medical education towards practical training. To summarise, the legal documentation provided a comprehensive overview of the formal structure of teaching and assessment; however, it did not offer insights into the rationale behind the alterations to the content of the documents. Conversely, scientific journal articles provide this background information, given that they are publications in which experts express their views on whether formal changes are useful or not. Finally, interviews and participatory observation provide insight into the actual work being conducted in a Skills Lab, including their commodities, and the attitudes towards the formal structure and the scientific arguments compared to everyday teaching and learning in these programmes.

Returning to the initial typology of Historical Institutionalism, and the analysis of transformation that the introduction of SBT and Skills Lab represent for medical education in Germany, two reference points of teaching and testing need to be assessed separately. *Teaching represents Layering, because, despite astonishment on the part of lecturers about disagreements between individual faculties, HI provides a solid explanation for the stickiness of singular faculties' slow transformation*. However, this transformation could only be considered a Displacement overall if faculties not implementing SBT showed a difference in study programme, but no direct line of argument towards regular or model study programme could be found. Both types have adopted and implemented SBT to varying degree, depending on the individual faculty. Another possibility of Displacement would be if a new examination order would implement SBT as a mandatory part of the state examination, by introducing SPs as assessment tools or OSCEs as assessment procedures for example. Faculties would also use SBT within their teaching programs with a high degree of certainty in that case to prepare their students for the state examination and, therefore, would also train their clinical teachers even more than under the Layering regime towards SBT.

However, SBT represent a Displacement when considering the individual faculties and their examination processes. This is because the practical exam formats, OSCEs in particular but also increasingly EPAs, take up resources and have downsized other formats of examination such as multiple-choice exams, which used to have more importance in the internal examination processes. Commodities and personnel seem almost irreplaceable in the conducting of such practical examinations, due to the fact that OSCEs take place at their location's Skills Labs in almost every case.

There are certain factors that need to be considered to assess possible future developments. First the funding of SBT. Skills Labs are a cost-intensive endeavour for any medical faculty, due to its personnel, and commodities, like simulators and SPs. It is noteworthy that in some cases commodities, especially "high tech" ones, often remain unused, such as the high-fidelity manikins or VR applications for example. It remains questionable whether the funding enthusiasm of the early 2010s, an enthusiasm which is also visible in the conducting of the NKLM, will subside and will ensure willingness to invest; however, the continuation of the programmes mentioned seems to be almost universally successful. Another factor stabilising this funding could be the high (and still increasing) number of students at medical faculties.

Another issue for SBT is medical education's (not yet entirely stabilised) competence-orientation. Two reasons important to understanding the difficulty of implementing competence as the leading didactical concept. Firstly, setting standards is not trivial, due to the need to both define and identify the necessary skills with which modern medical professionals, in an everchanging medical health care system, should be provided. For this reason, the work on the NKLM is an everchanging process, with numerous educators discussing and rearranging skills and competences within the different medical specialties and case compositions.

Where the definition of skills is given, they nevertheless need to be assessed, which is not an easy undertaking. Due to the general concept of competence being comprised of knowledge, skill and values, at least the latter is a latent character trait that is hard to grasp in formative assessment. However, the introduction of competence-orientation can be considered as both a cognitive (didactic) and social (pressure on healthcare system) claim (Abbott, 1988, pp. 57–66) by educators towards the necessary reforming of the system regarding education's overall development in Europe towards competence (Klieme et al., 2008). Furthermore, a German mixed-methods study about the transfer of knowledge in the clinical context revealed that experienced clinicians also withhold feedback, representing their experience, towards the novices to let them learn for themselves, while at the same time assessing how big possible damage (in the case of a mistake) could be (Wilkesmann et al., 2019, pp. 164–165). This phenomenon, observed in the clinical day-to-day workplace, aligns closely with the understanding of supervision and trust encapsulated by the EPAs.

The transformation of medical education is worthy of consideration in terms of the development of the medical profession overall, and the transformation is visible, fundamental and unfinished. *Whether and how SBT will be fully implemented as a leading concept within medical education or if it will continue to run underneath general conceptions of competence remains open at the state of things for this study; alternatively, funding and other pressure exerted onto the health care and educational systems might even evoke an entire rearrangement in the coming decades. This notwithstanding, medical education overall, inter alia with the introduction of SBT, has risen in its claims as a subordinated profession towards the medical profession through its professionalisation.*

7. Conclusion and Outlook

This study's final chapter aims to bring the different threads of analysis undertaken so far together. In this way, the central concepts and issues will be re-established, while the broader context will be opened up. *This study's objective was to trace developments in medical education as a subarea of medicine and, therefore, to characterise the transformation in terms of its professionalisation and institutionalisation. In this regard, SBT is understood as a placeholder for effects and transformations that are worth considering as important historical and contemporary changes relating to education, science and technology studies and to the medical profession.* At the same time, SBT is inevitably, through its didactical concepts, institutional composition and material constitution of commodities, considered to be bordering on all of the aforementioned discourses, thereby underlining this study's interdisciplinarity. At the same time, this study must be understood as primarily sociological, even as it introduces historical context and offers a diversification, less towards methodology and more towards the data driving the multi-layered analysis provided.

Professionalised SBT

The medical education offered by SBT is best understood by considering its foundation in reformist education, particularly in problem-based learning and competence-oriented education. Competence-oriented education has gained significant traction across various educational disciplines (Klieme et al., 2008). Consequently, it is not surprising that this shift has also affected medical education (Miani et al., 2015, p. 101). Nevertheless, the present study has demonstrated that the discourse surrounding such concepts predates their integration into curricula and institutions, as evidenced by the significant rise of Skills Labs at German medical faculties since the 2000s. There were also arguments for reformist education, international examples and pioneers in the German context from at least the 1980s onwards.

Moreover, it is important to consider that the constitution of SBT itself has undergone transformations throughout the specified time-frame from the 1980s to the 2020s. This is because their institutional and architectural representation has historically transitioned from the clinical environment that they seek to represent to a distinct form of their own, one characterised by the presence of modern rooms and equipment. This transition is exemplified by the German case studies conducted in North Rhine-Westphalia (i.e., initially sharing rooms with subspecialties to inhabitation of a self-designed, newly built facilities). Similar incremental changes have been observed in other Skills Lab facilities with long-standing and esteemed reputations, such as the role-model in Maastricht, Netherlands (Nott, 2023, p. 101).

The framework of Historical Institutionalism (HI) is an effective means of comprehending two aspects of a given phenomenon. Firstly, SBT was established earlier than the institutions of Skills Labs. Secondly, incremental changes occurred throughout SBT's rise. Accordingly, the present study focuses on stakeholders' options for action against the regime of the given institutions (Streeck & Thelen, 2005, p. 14). HI does not prioritise agents or structure, but rather conceptualises both as inherently interdependent (Streeck & Thelen, 2005, p. 19). This approach stands in contrast to other institutionalisms. It also does not restrict its perspective to critical junctures exclusively, because while these are important to observe, the majority of transformations, especially when internal, have a longer duration than immediate turmoil. The NKLM serves as a prime example of this dynamic, as it represents a critical juncture in the field's evolution, but it also exhibits incremental change (i.e., in the form of the GMA Committee for Practical Skills' work).

As medical education has become professionalised, oftentimes with regard to the abstract knowledge relating to didactical concepts such as competence or problem-based learning in Maastricht's case (Rotgans, 2012), so too have SBT and Skills Labs. This professionalisation is evidenced by the relocation of buildings, the increase in commodities, such as SP programmes and simulators, the expansion of staff, including permanently employed educators at the Skills Labs, the introduction of peer-to-peer teachers and the growth in subspecialties taught and assessed at the Skills Labs – as demonstrated in the NRW case studies and in accordance with international examples (Nott, 2023, p. 114).

However, concerns have been raised about the risks of SBT if it is not carried out properly. For instance, students may rigidly follow the scripts provided for 'effective communication', thereby ignoring their instincts when interacting with simulation patients (Chen, 2011). The reduced time available for interactions between educators and students could also hinder the development of model and moral learning in novice medical professionals, which is particularly necessary for SBT (Hardy, 2017). In addition, its artificial nature and the potential for the further biomedicalisation of medical education have been identified as areas of concern (Johnson, 2007; Nott & Harris, 2023; Prentice, 2013). Such criticism reflects the different cultures of medical education, and of education overall, changing the outlook on "the knowing bodies of future practitioners and the known bodies of their future patients [...]" (Nott, 2023, p. 101), which are learnt within medical studies. SBT, and its commodities of simulators, can also be understood through the lens of contemporary medicine's trust in technology and in its advancement.

When considering the relationship between technology and medical professionals, it becomes apparent that the clinical environment, whether constructed in Skills Labs or elsewhere, has to look, feel and be taught differently than it did several decades ago when most contemporary technology had not yet been introduced or implemented on such a wide scale; this is due to the computerisation of society as a whole (Greene, 2022). This has led to fundamental changes in professional practice, which is why the analysis of commodities can be considered to be at the forefront more than at any other time in the past (Greene, 2022;

Schubert, 2006, p. 81). Nevertheless, the persistence of the old must not be overlooked when analysing technological innovation (Edgerton, 2008).

In this regard, Abbott's theory of commodities is exceptionally useful, due to the notion that the technology in question also represents expert knowledge that has been materialised (Abbott, 1991). Novices, like medical students, initially struggle to integrate technology into their actions. They focus on the technology because it reduces and structures options, while experts can pay attention to other influences. This interdependence is also referenced in medical education, because the professionals have to continue to work and educate, through formulating individual belief systems towards technology, in technology's evolving and convoluted fields (i.e., devaluing the influence of technology itself without a didactical concept (see Chapter 5.3.), by finding workarounds and practicing bricolage in everyday education (ibid), if commodities are unreliable and creating categories of quality to assess the modifications of technology available (2013-V30-I1 (4), P. 2–3).

Abbott's account of the professionalisation of professional groups, medical education through SBT as a subordinate to medicine in this case, is useful for organising this analysis. Professional jurisdiction, as the struggle for dominance within a professional field through the application of abstract academic knowledge in professional practices in diagnosis, inference and treatment, can help differentiate between the different kinds of knowledge. In Abbott's view (1988), it can help to consider the different kinds of claims that are put forward. Furthermore, the whole study is outlined in terms of three jurisdictional arenas, providing structure and keeping the focus on specific aspects of professional work and on its transformations.

In both HI's and Abbott's account, the stakeholders at different levels of aggregation are understood as competitors in regard to the proclaimed enhancements. However, HI would not offer a direct analysis for the simultaneity of experts taking stances as individuals and representatives of an institution, while the mutual influence of the professional arenas does account for such instances. Furthermore, HI does not provide a term for the materiality of institutions or expert knowledge. In contrast, the commodities, as defined by Abbott (1991), are central to SBT and are a useful concept for professionalisation.

The embodiment of knowledge is not only inherent in the commodities, but also in their usage by students. Embodied learning has a long-standing history of realising and memorising through body movement, when engaging with maps representing geographical knowledge for example (i.e. tracing the objects on a map with the finger helps to create a sense of size and to focus on specific details of the map (Eddy, 2023, pp. 228–231)). Therefore, it seems no accident that the remembrance and sensory understanding of a body is commonly referred to as mapping and students create a map in their minds in practical medical education.

Guarrasi describes "how the standardised patient's body is transformed into an embodied map that systematises, organises, and classifies novices' knowledge into a unified system (...) the high-fidelity simulator is a valid educational tool not because it realistically mimics the body of the patient, but because it authentically represents the body of the patient as experienced by the medical professional" (Guarrasi, 2022, pp. 211–212). While some authors criticise this standardised body as a learning concept (Nott & Harris, 2023), others point to education as a 'smoothed out' version of practical skills, wherein going into the real clinical environment for the first time is as anxiety-inducing as it is organised through issues beyond the examination's singular steps (Duvivier et al., 2012, p. 347). To paraphrase Prentice, the commodities of SBT do not have to 'feel right'; they just need to be good enough for students to learn the basic bodily posture to work on and with them (Prentice, 2013, p. 233), while also warranting that a medical professional's abilities go far beyond the hands-on skills or other mechanistic abilities, which are not represented on a simulator (Prentice, 2013, p. 252).

Nonetheless, Skills Labs are neither the be all and end all of education for them, nor are they for medical education in general, as the educators have put it in so many words within the given expert interviews. Rather, they are a means to further the educational system's transformation (i.e., the implementation of competence-oriented concepts) and to face the transformation of the health care system in which medical education is embedded.

German Health Care Under Pressure

The German healthcare system faces many pressures to maintain its statutory health care system. One such factor is the shift in population composition, characterised by an ageing demographic, an escalation in chronic conditions and care dependency and a decline in fertility, thereby disrupting the generational contract. While this challenge is undoubtedly severe, it is not exclusive to Germany; rather, it is prevalent throughout most of the Western hemisphere (Busse, Blümel, Knieps, & Bärnighausen, 2017, p. 883). Consequently, the reduction of costs within the German healthcare system has been a recurring subject of debate since the mid-20th century (Illing, 2022; Janda, 2022, pp. 35–36). A change that originated in such austerity measures include diagnosis-related-groups, short DRGs, which will be considered in more detail below, due to the fact that the DRGs illustrate the pressure on the health care system with regard to the duration of patients' hospitalisation and with it the manner in which patients can be observed and can, thus, be available for students to learn from in the clinical context.

The prevailing DRGs system is predicated on the principle of categorising patients and their symptoms into billing groups. This system's overarching objective is to reduce the time that patients spend in hospital, thereby reducing costs and encouraging competition among hospitals to enhance quality (Braun, 2014, p. 91). The DRGs were introduced in Germany in 2004 for the purposes of patient care. A similar system of per-case rates for doctors had been introduced in 1996 as part of a broader programme of organisational reform (Braun, 2014, p. 92). A comparison of the statistical means of patients staying in German hospitals reveals a consistent downward trend in the number of days spent in hospital. In 1991, the average length of hospitalisation for patients was 14 days. This figure decreased from 8.9 to 7.7 days between 2003 and 2011, a development often cited as evidence of the phenomenon of 'bloody releases' of patients who had not yet recovered (Braun, 2014, p. 95). In principle, the DRGs should have increased the number of post-hospital care services and the overall number of physicians. However, surveys conducted during the specified timeframe contradict this hypothesis, suggesting an ineffective transition in healthcare logic (Braun, 2014, p. 96). Nevertheless, the introduction of DRGs is frequently regarded as a success story. This raises the question of why there is a discrepancy between the perception of change and the reality of hospital care. This phenomenon can be attributed to the reform inertia of institutions, which perpetuate the 'old reality' of universal health care and care guarantees rooted in the professional-ethical self-image, as opposed to the 'new reality' of profit orientation and the interpretation of patients as clients (Braun, 2014, p. 100). Braun's analysis offers a critical evaluation of the DRGs, concluding that they did not fulfil the reform potential anticipated with their introduction. Rather, the study identifies numerous critical, unintended consequences, including a deterioration in the quality of healthcare provided in German hospitals directly through per-case flat rates and DRGs (Braun 2014, p. 110, translation J.D.).

A study undertaken in 2009 concluded that the time constraints inherent to DRGs resulted in a diminution of the emphasis placed on knowledge transfer, particularly in the context of the hands-on teaching of novices within the clinical setting by clinical staff. This phenomenon was attributed to the fact that the additional time required for training was not adequately represented in clinical staff's working hours, making teaching unattractive (Wilkesmann, 2009, p. 267). This linkage of healthcare and education is also extended towards research with regard to university medicine, the central institutional composition of this study, for which university medicine is regarded as a leader. A recent study has indicated that the path-dependency of medical faculties in Germany frequently stands in opposition to the underlying rationale of governmental reforms of the health care system. This has the effect of negating both aspirations (i.e., for a transdisciplinary approach) and for greater interlinkages between healthcare and research, due to a lack of financial resources and the strong disciplinary and professional dominance of the medical faculties (Starystach & Dettmer,

2023). In order to comprehend the profound interconnection between medicine as a profession and medicine as a science, one has to consider the historical evolution of their relationship (P. H. Roth, 2021) and, as underlined previously, the emphasis is placed on the self-organisation of medical professionals; this should not be surprising either in terms of professional theory (Abbott, 1988) or with regard to German health care system's historical development (Blümel et al., 2020, pp. 29–30).

During the course of my on-site visits, healthcare professionals involved in teaching directly referred to the DRGs as problematic. This was due to the shortening of the time allotted for students to observe a patient's case, as well as the effect that a surgery that could address numerous problems at once can only be reimbursed by one specialty. Therefore, a priority must be proclaimed, leaving other specialties to work pro bono. Alternatively, several surgical procedures were scheduled to align with the DRG logic, thereby resulting in students observing fewer complex surgeries and, more significantly, increased stress for the patient. Therefore, from the standpoint of the SBT educators, it is preferable to provide an opportunity to practise clinical skills with SPs and simulators in SBT than it is to insufficiently observe diverse health conditions and to experience clinical work throughout their studies.

DRGs are just one example of the economic pressure in the German healthcare system, with medical staff adopting various coping strategies ranging from cooperation to silent resistance (Molzberger, 2020). DRGs are also not an exclusively German phenomenon, but have their origins in the Anglo-American context. Hyheyoung Oh Nelson describes the influence of DRGs in the US context as a 'hidden curriculum' for the decision-making process in patient management and for the training of medical newcomers (Oh Nelson, 2022). As a result, medical professionals are constantly weighing up the economic, legal and professional reasons to treat a patient. This becomes even more complex when considering the inter-professional organisation of hospitals and the influence of private companies on medical commodities and infrastructure (Oh Nelson, 2022, p. 40).

The final two factors, which could be considered for further study in the field of medical education, also exist to a varying degree in the

German context. German hospitals are complex organisations, due to their interprofessional composition (Wilkesmann, 2009). In relation to the issue of working experience in the clinical context across diverse healthcare systems, Germany has exhibited a consistent decline in the number of patients under the care of doctors and this juxtaposed with a parallel rise in the provision of care by nurses and by other care providers. Notably, Germany has the highest work intensity for nursing staff among all of the OECD countries (Braun, 2014, p. 93). This notwithstanding, the privatisation trend is less pronounced than it is in the US, but it is also gaining momentum in the German healthcare system (Blümel et al., 2020, pp. 29–30). These findings put pressure on the medical educational system as a subpart of the medical system because they have to evolve and – as shown – adapt through professionalisation to the new circumstances.

Ongoing Fight for Turf – A Fight for Education

A 2016 article by Stosch and Schnabel, who were then chairmen of the GMA Committee Practical Skills, is entitled 'Practical Skills en Route to Professionalism' (2016-V33-I4 (13)). As examined previously, this article's title and content traces the teaching of medical skills back to SBT's introduction in Germany in the 1990s and can be understood as agenda-setting (see Chapter 6). Nevertheless, the transformation described can be understood as a Layering of teaching methods and a Displacement of assessment methods employed in contemporary medical education in Germany. However, the issue of a new examination order remains unresolved, thereby limiting the options for medical education's further professionalisation, specifically for SBT, to speculative considerations within the legal arena and diverse public opinions.

Furthermore, one of the many current and continuous debates in contemporary medical education concerns the number of students. The demographic change is a considerable concern for the health care system in Germany, as with many European countries. The danger of a big generation of doctors being retired, but not enough young doctors coming afterwards to fill these gaps, is real and constituent for the state system. What is interesting for the MES is that different experts propose different solutions. While the BÄK asks for a rise in student numbers (Bundesärztekammer [BÄK], 2023, p. 31), experts like the current director of the MFT disagree, identifying the problem elsewhere. The student numbers have risen steadily in universities, but the contents and organisation of the study programmes need to be doable for the medical faculties. He hints at the – still ongoing – debate about a reform of the examination order as the direction that the future education should take (Gerlof, 2023). The need for a reform is neither new, since the new examination order was announced in 2018, nor is the MFT the only institution that asks for a final decision to come swiftly. For example, the Hessian Landesärztekammer is, together with an initiative from the medical student organisation, summarising and criticising the missing progress, due to the COVID-19 crisis and a change of government that often cause delays do the "inherited" topics and collide with initiatives from a new ministry line-up (Hessische Landesärztekammer [Hessian LÄK), 2024).

The IMPP was heavily involved in the examination's development. While the competence orientation was welcomed by the IMPP, the examinations were debated as being too detail-oriented. The use of multiple-choice questions as a tool was often debated, as was what should be covered in the NKLM. IMPP reforms received fierce criticism, including from students, who are represented by the „Bundesvertretung der Medizinstudierenden in Deutschland" (Bundesvertretung der Medizinstudierenden in Deutschland [bvmd], 2019). In 2021, Jana Jünger was suspended and replaced by Jan Carl Becker. Becker himself had a strong focus on medical education, including a Master of Medical Education from Bern, Switzerland (not the German programme) and worked at the University of Münster at the „Institut für Ausbildung und Studienangelegenheiten (IfAS)". The institute initially founded by Dietrich Habeck. Becker was also involved in the development of the „Masterplan Medizin 2020". Nevertheless, supporters of Jünger feared that the new examination order would come to a standstill with this change of director. However, while the Federal Ministry of Health still held out for the new examination order, further changes were being made to the examination organisation (Ärzteblatt, 2021; Institut für medizinische und pharmazeutische Prüfungsfragen [IMPP], 2022).

A recent press release from the IMPP stated that the transfer of old MC questions to commercial publishers would be stopped immediately (IMPP, 2024). This was the IMPPs response to the fact that the exam questions became more detailed in recent years and were, thus, more remote from everyday life. By retaining questions, students can no longer learn with questions taken from actual test sheets. Still, some student councils of medical students, for example in Mannheim, welcomed the IMPPs decision (Fachschaft Medizin Mannheim, 2024). Thieme via medici, one of the major commercial publishers, also welcomed the decision, adding that they also want to provide a realistic learning experience, pointing to the immense number of questions from previous years that are still available to learn with when using Thieme via medici materials (Thieme via medici, 2024).

Meanwhile, the NKLM remains unfinished. Educators are working on it and the MFT has declared it mandatory for faculties in 2024. Consequently, they elect to establish their own regulations, despite the ministry's decision to refrain from issuing a new examination order (Medizinischer Fakultätentag (MFT) Deutschland, 2024, p. 193; Medizinischer Fakultätentag (MFT) Deutschland, & Deutsche Hochschulmedizin e.V., 2023).

The German educational reforms cannot be understood in isolation. The German system is constantly compared to international constructions, especially in terms of competence-orientation (Jünger, 2018; Wissenschaftsrat, 2018, pp. 19–20). As shown previously, educational experts have demanded competence-orientation in progressive education since the 1980s, comparing it to international standards (Renschler, 1987). A 2017 study showed that successful reform efforts are based on general policy-making and can favour either academic, abstract education or workplace, practical, problem-oriented outlines (Afdal, 2017). This study showed how specific actors, be it individuals or associations, try to influence the general policy through voicing their stances in the three jurisdictional arenas mentioned regarding medical education.

However, an incorrect conclusion regarding Abbott's theory and the empirical case of SBT could be that a greater focus on practical elements in medical training would be contrary to the high proportion of abstract knowledge (1988). This can be found in the motto: 'To the things and patients, away from the textbooks'. However, this conclusion is wrong on two counts: in terms of both the empirical case and for the theoretical basis. Firstly, it is clear that abolishing students' theoretical knowledge would not make sense for teachers, including those in Skills Labs and in other simulation training courses, because it is precisely the testing of skills that relates to training in abstract knowledge. Practical or communicative skills are nondescript without the specialist knowledge and should ideally not pass as correct behaviour in a simulation.

Secondly, Abbotts's theory (1988) defines professional behaviour as a complex process of recognising problems, weighing up the various solutions and then implementing the solution (diagnosis, inference and treatment). This means that simulation training courses can logically only work *with* abstract knowledge and never *against* abstract knowledge, otherwise they would simply be unprofessional. Nevertheless, it is possible for either the simulations to be poorly organised in teaching practice or for the results to be difficult to verify with regard to abstract knowledge. For example, a student may perform actions blindly without understanding why. However, this is usually prevented by appropriate knowledge checks and a poorly executed method in an isolated case does not undermine the training objective as a whole. The key issue here is the interaction between the most recent scientific discoveries and the process of teaching and thinking about how to translate these scientific claims into clear, understandable lessons (Abbott, 1988, p. 72).

Discussion and Outlook: Professionalisation of Health Care Education

This study's final subchapter aims to facilitate a critical discussion of possible extensions to its methodological and theoretical composition, thereby offering an extension to the research about the professionalisation of healthcare education.

One extension of this study would be to include direct critiques of SBT. While the JME journal articles (data type II) and expert interviewees and participatory observations (data type III) offer arguments for and against the use of SBT in medical education and the establishment of Skills Labs as institutions, no explicit critique of SBT was sought out. This could be achieved by interviewing clinical lecturers who feel that the teaching and testing procedures at the Skills Labs are inadequate. However, it could be difficult to choose a group of clinical teachers or medical subspecialty to act as representatives of organised resistance to SBT, and it would be necessary to ensure that they are familiar with SBT's didactic concepts, since while the Skills Lab personnel can be considered SBT experts, the clinical lecturers may not be.

Another extension of the given study concerns the analysis of the transformation undertaken within this study, which was not aimed towards all medical faculties in Germany and their implementation of Skills Labs. The Skills Labs should have been surveyed directly in order to achieve this aim, as undertaken by Fandler several years ago (Fandler, 2012). The description of incremental transformation aimed for within this study was much less the initial implementation or narrations of frontrunners of the given field, but the changes leading up to (i.e., the decision to found) a Skills Lab or how the institutions evolved regarding room organisation and commodities. The smaller scale of focusing on North Rhine-Westphalia made it possible to retrace such changes more closely. One can assume that similar changes have occurred in all of Germany, since the studies of overall prominence suggest accordingly (Fandler, 2012; Segarra et al., 2008; St. Pierre & Breuer, 2018), but further comparative research could underline the SBT's founding stages of implementation in Germany.

The final beneficial extension of the given study could involve the commercial aspect of SBT commodities, with a focus on the impact of manufacturers. While this was beyond the scope of the present study, the topic is frequently discussed at medical education conferences and undoubtedly informs educators' decisions regarding the acquisition, maintenance and repair of equipment in Skills Labs. This focus would also highlight the importance of social mechanisms beyond medicine and medical education, such as globalisation and neoliberalism (Evetts, 2008).

This study has provided a detailed examination of the transformation of medical education, using the rise of SBT as an example, by applying complex reasoning as defined by Creswell (Creswell, 2013, p. 45). This multi-contextualised approach was achieved through directed and summative content analysis (Hsieh & Shannon, 2005, p. 1286), was based on the three professional arenas and on their respective data. This iterative analysis, alongside the second theory of institutional transformation, enables the study's methodology to achieve more robust intersubjectivity in the findings and for sufficient triangulation (Fielding & Fielding, 1986, p. 10). The study offers insights into diverse theoretical and empirical fields of inquiry: the sociology of profession, of medicine, of education, but also the historical transformation of education, of medicine, and institutions as well as inspirations from science and technology studies.

A subject that was mentioned and should not be overlooked, particularly given that this study was based on a theory of professions, are the boundaries of the profession under discussion. Whilst medical education can be considered within the, seemingly well-established and almost hardened, position of medicine as both a discipline and a profession, there are other professional fields in proximity to medicine that have undergone processes of professionalisation and academisation, which were not included in this study. Consequently, further research is recommended, given that they encounter analogous challenges and issues in their education, such as the academisation of nursing science and midwifery in Germany. Initial research has been conducted on the compatibility of medicine with other fields of study (Sander, 2023) and the trend within these disciplines to utilise SBT in their educational curricula has also been documented (Akselbo & Aune, 2023). In this context, it would be beneficial to employ Abbotts's theoretical framework, particularly his examination of various professional settlements, to analyse the relationship between medicine and the other professions under discussion (Abbott, 1988, p. 86).

Adopting this perspective on interprofessional work in healthcare systems and education provides a clearer picture of how medical categories of expertise, knowledge and expectations evolve in response to political and societal demands (Jutel & Nettleton, 2011, p. 793; Nettleton, 2021, p. 27) and highlights the importance of diverse educational approaches (Nott et al., 2022). Such research recognises future healthcare professionals as individuals who must navigate an ever-changing and complex system (Molzberger, 2020). Consequently, this study has examined didactic theorists' claims in favour of robust mentoring systems that foster professional competence as a combination of knowledge, skills and values (Epstein & Hundert, 2002, p. 233; Fabry, 2022).

In conclusion, it is evident that the fundamental argument of Abbott's theoretical framework is founded upon the concept of competition (1988). This signifies that no single profession can be regarded as an isolated entity, devoid of its societal context or as a coherent whole. Instead, it is characterised by the presence of subfields and paradoxes, both within and beyond its immediate domain, which necessitate protection and defence due to their potential for contestation. This theory of negotiation is distinct from traditional market theories that rely on either supply and demand or the accumulation of social power, seemingly out of nowhere. Instead, it is characterised by the dynamics of cooperation and subordination, which are facilitated by modern societal systems, including legal frameworks, public discourse and specific working institutions and environments. Consequently, only the combination of historical and contemporary insight into these negotiations can illuminate a professional field's incremental evolution, such as medical education and the rise of SBT as a methodology (Abbott, 1988, p. 2; Streeck & Thelen, 2005).

8. Appendix – Available Digitally

The Appendix, which is available digitally, invites readers to delve deeper into the procedures and nature of the documents and original materials, especially the expert interviews with SBT educators (Appendix 6), used in this book.

- Appendix 1: State universities with medical departments in Germany (as of 2025)
- Appendix 2: Examination Orders and Curricula analysed
- Appendix 3: GMS Journal of Medical Education Articles analysed
- Appendix 4: Interview Guide
- Appendix 5: Selected findings with regard to the public arena
- Appendix 6: Transcript of Interview sections quoted
- Appendix 7: Detailed analytical procedure
- Appendix 8: Transformation of SBT in North Rhine-Westphalia

Available at https://tvl.ink/9783839478349

References

Aagaard, K. (2017). The Evolution of a National Research Funding System: Transformative Change Through Layering and Displacement. Minerva, 55(3), 279–297. https://doi.org/10.1007/s11024-017-9317-1

Abbott, A. (1988). The System of Professions. An Essay on the Division of Expert Labour. University of Chicago Press.

Abbott, A. (1991). The Future of Professions. Occupation and Expertise in the Age of Organisation. Research in the Sociology of Organisations, 8, 17–42.

Abbott, A. (2001). Chaos of disciplines. Univ. of Chicago Press.

Abbott, A. (2020). Zeit zählt: Grundzüge einer prozessualen Soziologie. Hamburger Edition.

Afdal, H. W. (2017). "Research-based" and "profession-oriented" as prominent knowledge discourses in curriculum restructuring of professional programs. Higher Education, 74(3), 401–418. https://doi.org/10.1007/s10734-016-9998-7

Agha, R. A., & Fowler, A. J. (2015). The role and validity of surgical simulation. International Surgery, 100(2), 350–357. https://doi.org/10.9738/INTSURG-D-14-00004.1

Ahlers, O. (2018). Der richtige Rahmen entscheidet: curriculare Implementierung der Simulation. In M. St. Pierre & G. Breuer (Eds.), Simulation in der Medizin (pp. 91–96). Springer.

Akselbo, I., & Aune, I. (2023). How Can we Use Simulation to Improve Competencies in Nursing? Springer International Publishing. https://doi.org/10.1007/978-3-031-10399-5

Armstrong, D. (1983). Political anatomy of the body: Medical knowledge in Britain in the twentieth century (1. publ). Cambridge Univ. Pr.

Ärzteblatt (2021, July 6). IMPP-Direktorin Jünger offenbar vorläufig suspendiert. Ärzteblatt. https://www.aerzteblatt.de/nachrichten/125264/IMPP-Direktorin-Juenger-offenbar-vorlaeufig-suspendiert

Atzeni, G. (2016). Professionelles Erwartungsmanagement: Zur soziologischen Bedeutung der Sozialfigur Arzt (1. Auflage). Nomos.

Atzl, I., Helms, R., Neuner, S., & Schilling, R. (2013). Praxiswelten. Zur Geschichte der Begegnungen von Arzt und Patient. druckpruskil.

Backlund, P., Maurin Söderholm, H., Engström, H., Andersson Hagiwara, M., & Lebram, M. (2018). Breaking Out of the Bubble Putting Simulation Into Context to Increase Immersion and Performance. Simulation & Gaming, 49(6), 642–660. https://doi.org/10.1177/1046878118772612

Barrows, H. S. (1968). Simulated patients in medical teaching. Canadian Medical Association Journal, 98(14), 674–676.

Becker, H. S., Strauss, A., Hughes, E. C., & Geer, B. (2007). Boys in white: Student culture in medical school (1. paperback ed., 10. print). Transaction Publ.

Beese, C. (2024). Bauen für die Wissenschaft. Zum Entstehungsprozess des anatomischen Theaters in Göttingen zwischen 1733 und 1737. In F. Abe & C. Beese (Eds.), Bauten – Bilder – Geschichten. Kunsthistorische Perspektiven auf Architektur (pp. 219–236). arthistoricum.net.

Blohm, M., Lauter, J., Branchereau, S., Krautter, M., Köhl-Hackert, N., Jünger, J [J.], Herzog, W., & Nikendei, C. (2015). "Peer-assisted learning" (PAL) in the Skills-Lab—an inventory at the medical faculties of the Federal Republic of Germany. GMS Zeitschrift Fur Medizinische Ausbildung, 32(1), Doc10. https://doi.org/10.3205/zma000952

Blümel, M., Spranger, A., & et al. (2020). Germany: health system review. i – 273.: European Observatory on Health Systems and Policies.

Blyth, M. (2002). Great Transformations: Economic Ideas and Institutional Change in the Twentieth Century (1st ed.). Cambridge University Press. https://ebookcentral.proquest.com/lib/kxp/detail.action?docID=3004707

Bowyer, M. W., Streete, K. A., Muniz, G. M., & Liu, A. V. (2008). Immersive Virtual Environments for Medical Training. Seminars in Colon and Rectal Surgery, 19(2), 90–97. https://doi.org/10.1053/j.scrs.2008.02.005

Braun, B. (2014). Auswirkungen der DRGs auf Versorgungsqualität und Arbeitsbedingungen im Krankenhaus. In A. Manzei & R. Schmiede (Eds.), 20 Jahre Wettbewerb im Gesundheitswesen (pp. 91–113). Springer Fachmedien Wiesbaden.

Brown, W. J., & Tortorella, R. A. W. (2020). Hybrid medical simulation – a systematic literature review. Smart Learning Environments, 7(1). https://doi.org/10.1186/s40561-020-00127-6

Buck, G. H. (1991). Development of simulators in medical education. Gesnerus, 48 Pt 1, 7–28.

Bugaj, T. J., & Nikendei, C. (2016). Practical Clinical Training in Skills Labs: Theory and Practice. GMS Journal for Medical Education, 33(4). https://doi.org/10.3205/zma001062

Bundesärztekammer. (2023). Tätigkeitsbericht 2022 der Bundesärztekammer.

Bundesvertretung der Medizinstudierenden in Deutschland. (2019, November 18). Übereilte Anpassungen der medizinischen Staatsexamina [Press release]. https://www.bvmd.de/wp-content/uploads/2022/01/Offener_Brief_der_bvmd_an_das_impp.pdf

Busse, R., Blümel, M., Knieps, F., & Bärnighausen, T. (2017). Statutory health insurance in Germany: A health system shaped by 135 years of solidarity, self-governance, and competition. Lancet (London, England), 390(10097), 882–897. https://doi.org/10.1016/S0140-6736(17)31280-1

Busse, R., Blümel, M., & Spranger, A. (2017). Das deutsche Gesundheitssystem: Akteure, Daten, Analysen (2. Auflage). MWV Medizinisch Wissenschaftliche Verlagsgesellschaft. https://doi.org/10.32745/9783954662623

Casper, M. J., & Morrison, D. R. (2010). Medical sociology and technology: Critical engagements. Journal of Health and Social Behavior, 51, 120–132. https://doi.org/10.1177/0022146510383493

Cate, O. ten (2013). Nuts and bolts of entrustable professional activities. Journal of Graduate Medical Education, 5(1), 157–158. https://doi.org/10.4300/JGME-D-12-00380.1

Chen, R. P. (2011). Moral imagination in simulation-based communication skills training. Nursing Ethics, 18(1), 102–111. https://doi.org/10.1177/0969733010386163

Collier, D., & Munck, G. L. (Eds.). (2022). Critical junctures and historical legacies: Insights and methods for comparative social science. Rowman & Littlefield.

Collins, P., & Harden, R. M. (1998). AMEE Medical Education Guide No. 13: real patients, simulated patients and simulators in clinical examinations. Medical Teacher, 20(6), 508–521. https://doi.org/10.1080/01421599880210

Creswell, J. W. (2013). Qualitative inquiry and research design: Choosing among five approaches (Third edition). Sage.

Dick, M. (2016). Professionsentwicklung als Forschungs- und Handlungsfeld. In M. Dick, W. Marotzki, & H. A. Mieg (Eds.), Handbuch Professionsentwicklung (1. Auflage, pp. 9–24). UTB.

Duffin, J. (2021). History of medicine: A scandalously short introduction (Third edition). University of Toronto Press.

Duvivier, R. J., van Geel, K., van Dalen, J., Scherpbier, A. J. J. A., & van der Vleuten, C. P. M. (2012). Learning physical examination skills outside timetabled training sessions: What happens and why? Advances in Health Sciences Education : Theory and Practice, 17(3), 339–355. https://doi.org/10.1007/s10459-011-9312-5

Eckart, W. U. (2019). Anatomisches Theater. In F. Jaeger (Ed.), Enzyklopädie der Neuzeit Online. BRILL.

Eddy, M. D. (2023). Media & the mind: Art, science, and notebooks as paper machines, 1700–1830. The University of Chicago Press.

Edgerton, D. E. H. (2008). The shock of the old: Technology and global history since 1900 (Paperback ed.). Profile Books.

Epstein, R. M., & Hundert, E. M. (2002). Defining and assessing professional competence. JAMA, 287(2), 226–235. https://doi.org/10.1001/jama.287.2.226

Evetts, J. (2008). Introduction. Professional Work in Europe.: Concepts, theories, and methodologies. European Societies, 10(4), 525–544. https://doi.org/10.1080/14616690701871696

Evetts, J. (2013). Professionalism: Value and ideology. Current Sociology, 61(5-6), 778–796. https://doi.org/10.1177/0011392113479316

Fabry, G. (2022). Wie lassen sich professionelle Kompetenzen im Medizinstudium vermitteln? Ethik in Der Medizin, 34(3), 287–299. https://doi.org/10.1007/s00481-022-00695-w

Fachschaft Medizin Mannheim. (2024, January 29). FiMM kommentiert: Aktuelle Entwicklungen beim IMPP [Press release]. https://www.fimm-online.de/fimm-kommentiert-aktuelle-entwicklungen-beim-impp/

Fandler, M. (2012). Simulation in der medizinischen Ausbildung. Der zunehmende Fokus auf klinisch-praktische Fertigkeiten, Auf- und Ausbau von klinischen Simulations- und Trainingszentren an den Medizinischen Universitäten und Fakultäten in Österreich, Deutschland und der Schweiz.: Unveröffentlichte Diplomarbeit. https://online.medunigraz.at/mug_online/wbAbs.showThesis?pThesisNr=19700&pOrgNr=1&pPersNr=79271

Federal Ministry of Health, & German Research Foundation. (N.N.). About the portal German Medical Science. https://www.egms.de/static/en/about.htm

Fielding, N., & Fielding, J. L. (1986). Linking data. Sage.

Foucault, M. (1973). Wahnsinn und Gesellschaft: Eine Geschichte des Wahns im Zeitalter der Vernunft (U. Köppen, Trans.) (Erste Auflage). suhrkamp taschenbuch wissenschaft: Vol. 39. Suhrkamp.

Freidson, E. (1995). Profession of medicine: A study of the sociology of applied knowledge [Nachdr.]. Univ. of Chicago Press.

Freidson, E. (2001). Professionalism: The third logic (1. publ). Polity Press.

Geschäftsstelle Heinrich-Heine-Universität Düsseldorf. (2024). Homepage Landesakadmie für Medizinische Ausbildung (LAMA) NRW. https://www.medizin.hhu.de/lama/ueber-uns

Gesellschaft für Medizinische Ausbildung. (2021). Satzung der GMA Sept. 2021. Gesellschaft für Medizinische Ausbildung. https://gesellschaft-medizinische-ausbildung.org/gma/satzung.html

Gobo, G. (2023). Mixed Methods and Their Pragmatic Approach: Is There a Risk of Being Entangled in a Positivist Epistemology and Methodology? Limits, Pitfalls and Consequences of a Bricolage Methodology. Forum Qualitative Sozialforschung / Forum: Qualitative Social Research, 24(1).

Goffman, E. (2016). Asyle: Über die soziale Situation psychiatrischer Patienten und anderer Insassen (20. Auflage). Suhrkamp.

Görlich, D., & Friederichs, H. (2021). Using longitudinal progress test data to determine the effect size of learning in undergraduate medical education – a retrospective, single-center, mixed model analysis of progress testing results. Medical Education Online, 26(1). https://doi.org/10.1080/10872981.2021.1972505

Gradmann, C., & Simon, J. (Eds.). (2010). Evaluating and Standardizing Therapeutic Agents, 1890–1950. Palgrave Macmillan UK.

Greene, J. A. (2022). The doctor who wasn't there: Technology, history, and the limits of telehealth. The University of Chicago Press.

Greif, A., & Laitin, D. (2004). A Theory of Endogenous Institutional Change. APSR, 98(4), 633–652.

Guarrasi, I. (2022). Hearts: The Heart of the simulated matter: Interprofessional training and practices of clinical care. In J. Nott, A. Harris, R. Vaden Allison, H. van Lente, C. F. Sánchez Burmester, A. Wojcik, & S. Wyatt (Eds.), Making sense of medicine: Material culture and the reproduction of medical knowledge (pp. 205–218). Intellect.

Hallonsten, O., & Heinze, T [T.] (2013). From particle physics to photon science: Multi-dimensional and multi-level renewal at DESY and SLAC. Science and Public Policy, 40(5), 591–603. https://doi.org/10.1093/scipol/sct009

Harden, R. M. (1988). What is an OSCE? Medical Teacher, 10(1), 19–22. https://doi.org/10.3109/01421598809019321

Harden, R. M., Stevenson, M., Downie, W. W., & Wilson, G. M. (1975). Assessment of clinical competence using objective structured examination. British Medical Journal, 1(5955), 447–451. https://doi.org/10.1136/bmj.1.5955.447

Hardy, C. (2017). Empathizing with patients: The role of interaction and narratives in providing better patient care. Medicine, Health Care, and Philosophy, 20(2), 237–248. https://doi.org/10.1007/s11019-016-9746-x

Harris, A., & Harder, C. den. (2022). Balloons: Lessons from a balloon. In J. Nott, A. Harris, R. Vaden Allison, H. van Lente, C. F. Sánchez Burmester, A. Wojcik, & S. Wyatt (Eds.), Making sense of medicine: Material culture and the reproduction of medical knowledge (pp. 49–55). Intellect.

Heinze, T [Thomas], & Münch, R. (2012). Intellektuelle Erneuerung der Forschung durch institutionellen Wandel. In T. Heinze & G. Krücken (Eds.), Institutionelle Erneuerungsfähigkeit der Forschung (pp. 15–38). VS Verlag für Sozialwissenschaften. https://doi.org/10.1007/978-3-531-94274-2_2

Hessische Landesärztekammer [Hessian LÄK). (2024, February 16). Endlich Novellierung der Ärztlichen Approbationsordnung beschließen! https://www.bundesaerztekammer.de/presse/aktuelles/detail/endlich-novellierung-der-aerztlichen-approbationsordnung-beschliessen

Medizinischer Fakultätentag (MFT) Deutschland. (2009, May 22). Länderübergreifender Studiengang wird zum Erfolgsmodell – Akademie für Ausbildung in der Hochschulmedizin zieht Bilanz [Press release]. https://idw-online.de/de/news?print=1&id=316689

Hines, L. M. (2011). Mystery Solved: Detective Skills and the Historian's Craft. In S. D. Lapan, F. J. Riemer, & M. T. Quartaroli (Eds.), Qualitative Research: An Introduction to Methods and Designs (pp. 137–162). Wiley.

Hoebel, T., Knöbl, W., & Sahr, A. (2020). Reputation und Randständigkeit: Andrew Abbott und die Suche nach der prozessualen Soziologie. Andrew Abbott: Zeit zählt: Grundzüge einer prozessualen Soziologie. SSOAR, GESIS – Leibniz-Institut für Sozialwissenschaften e.V; Hamburger Edition.

Hsieh, H.-F., & Shannon, S. E. (2005). Three approaches to qualitative content analysis. Qualitative Health Research, 15(9), 1277–1288.

Hughes, E. C. (Ed.). (2015 – Orig 1958). Classic Reprint Series. Men and their work. Forgotten Books.

Huwendiek, S., & Haag, M. (2014). Der virtuelle Patient im Rahmen der medizinischen Ausbildung. In W. Niederlag, H. U. Lemke, H. Lehrach, & H.-O. Peitgen (Eds.), Der virtuelle Patient (2. edition, pp. 207–216). De Gruyter.

Ilgen, J. S., Sherbino, J., & Cook, D. A. (2013). Technology-enhanced simulation in emergency medicine: A systematic review and meta-analysis. Academic Emergency Medicine : Official Journal of the Society for Academic Emergency Medicine, 20(2), 117–127. https://doi.org/10.1111/acem.12076

Illing, F. (2022). Gesundheitspolitik in Deutschland: Eine Chronologie der Gesundheitsreformen der Bundesrepublik Deutschland (2. Auflage). Springer VS.

Institut für Ausbildung und Studienangelegenheiten. (N.N.). Organisation [Press release]. https://medicampus.uni-muenster.de/ifas.html

Institut für medizinische und pharmazeutische Prüfungsfragen. (2022, November 10). Jan Carl Becker ist neuer Direktor des Instituts für medizinische und pharmazeutische Prüfungsfragen [Press release]. https://www.impp.de/files/PDF/IMPP-News/IMPP-Medieninformation-Neuer-IMPP-Direktor-2022-11-10.pdf

Institut für medizinische und pharmazeutische Prüfungsfragen. (2024, January 26). Institut für medizinische und pharmazeutische Prüfungsfragen beendet kommerzielle Lizenzierung von Prüfungsaufgaben [Press release]. https://www.impp.de/files/PDF/IMPP-News/IMPP-Medieninformation-Lizenzierung-Pr%C3%BCfungsfragen-20240126.pdf

Issleib, M., Schmidt, G., Käser, M., & Breuer, G. (2018). Simulation trotz knapper Kassen: Finanzplanung. In M. St.Pierre & G. Breuer (Eds.), Simulation in der Medizin (pp. 61–71). Springer.

Jaki, C., St.Pierre, M., & Breuer, G. (2018). Vom Zimmer zum Zentrum – "form follows function". In M. St. Pierre & G. Breuer (Eds.), Simulation in der Medizin (pp. 21–47). Springer. https://doi.org/10.1007/978-3-662-54566-9_2

Janda, C. (2022). Medizinrecht (5., überarbeitete und aktualisierte Auflage). UVK Verlag.

Jappe, A., Pithan, D., & Heinze, T [Thomas] (2018). Does bibliometric research confer legitimacy to research assessment practice? A sociological study of reputational control, 1972–2016. PloS One, 13(6), e0199031. https://doi.org/10.1371/journal.pone.0199031

Johnson, E. (2007). Surgical simulators and simulated surgeons: Reconstituting medical practice and practitioners in simulations. Social Studies of Science, 37(4), 585–608. https://doi.org/10.1177/0306312706072179

Jones, F., Passos-Neto, C., & Melro Braghiroli, O. (2015). Simulation in Medical Education: Brief history and methodology. Principles and Practice of Clinical Research Journal, 1(2), 56–63. https://doi.org/10.21801/ppcrj.2015.12.8

Jünger, J [Jana] (2018). Kompetenzorientiert prüfen im Staatsexamen Medizin [Competence-based assessment in the national licensing examination in Germany]. Bundesgesundheitsblatt, Gesundheitsforschung, Gesundheitsschutz, 61(2), 171–177. https://doi.org/10.1007/s00103-017-2668-9

Jutel, A., & Nettleton, S. (2011). Towards a sociology of diagnosis: Reflections and opportunities. Social Science & Medicine (1982), 73(6), 793–800. https://doi.org/10.1016/j.socscimed.2011.07.014

Kamp, M. A., Malzkorn, B., Sass, C. von, DiMeco, F., Hadjipanayis, C. G., Senft, C., Rapp, M., Gepfner-Tuma, I., Fountas, K., Krieg, S. M., Neukirchen, M., Florian, I. Ș., Schnell, O., Mijderwijk, H.-J., Perin, A., Baumgarten, P., van Lieshout, J. H., Thon, N., Renovanz, M., . . . Sabel, M. (2021). Proposed definition of competencies for surgical neuro-oncology training. Journal of Neuro-Oncology, 153(1), 121–131. https://doi.org/10.1007/s11060-021-03750-6

Karsch, F. (2015). Medizin zwischen Markt und Moral: Zur Kommerzialisierung ärztlicher Handlungsfelder. Transcript.

Kelle, U., & Reith, F. (2023). Strangers in Paradigms!? Alternatives to Paradigm-Bound Methodology and Methodological Confessionalism. Forum Qualitative Sozialforschung / Forum: Qualitative Social Research, 24(1).

Kliche, O. (2015). Simulationspatienten in der medizinischen Ausbildung [Dissertation]. Universität zu Köln, Köln. Verlag für Gesprächsforschung Mannheim

Klieme, E., Hartig, J., & Rauch, D. (2008). The Concept of Competence in Educational Contexts. In J. Hartig, E. Klieme, & D. Deuter (Eds.), Assessment of Competencies in Educational Contexts (pp. 3–22). Hogrefe and Huber Publishers.

Knappertsbusch, F., Schreier, M., Burzan, N., & Fielding, N. (2023). Innovative Applications and Future Directions in Mixed Methods and Multimethod Social Research. Forum Qualitative Sozialforschung / Forum: Qualitative Social Research, 24(1).

Kneebone, R., & Oakes, F. (2022). Textiles: Materialities of Surgery: Learnig Through Thread. In J. Nott, A. Harris, R. Vaden Allison, H. van Lente, C. F. Sánchez Burmester, A. Wojcik, & S. Wyatt (Eds.), Making sense of medicine: Material culture and the reproduction of medical knowledge (pp. 394–403). Intellect.

Kögel, A. (2021). Medizin und Gesellschaft: Eine Einführung in die Medizinsoziologie (1. Auflage). Verlag W. Kohlhammer. http://www.kohlhammer.de/wms/instances/KOB/appDE/nav_product.php?product=978-3-17-037294-8

Krause, M. (2021). Model Cases. University of Chicago Press. https://doi.org/10.7208/chicago/9780226780979.001.0001

Kuckartz, U., & Rädiker, S. (2020). Fokussierte Interviewanalyse mit MAXQDA. Springer Fachmedien Wiesbaden. https://doi.org/10.1007/978-3-658-31468-2

Kühlmeyer, K., Muckel, P., & Breuer, F. (2020). Qualitative Inhaltsanalysen und Grounded-Theory-Methodologien im Vergleich.: Varianten und Profile der „Instruktionalität" qualitativer Auswertungsverfahren. Forum Qualitative Sozialforschung / Forum: Qualitative Social Research, 1(21).

Kurtz, S., Silverman, J., Draper, J., van Dalen, J., & Platt, F. W. (2017). Teaching and Learning Communication Skills in Medicine. CRC Press. https://doi.org/10.1201/9781315378398

Land, B. (2018). Das deutsche Gesundheitssystem – Struktur und Finanzierung: Wissen für Pflege- und Therapieberufe (1. Auflage). Verlag W. Kohlhammer.

Lareau, A. (2021). Listening to people: A practical guide to interviewing, participant observation, data analysis, and writing it all up. Chicago guides to writing, editing, and publishing. The University of Chicago Press.

Larson, M. S. (1977). The rise of professionalism: A sociological analysis. University of California Press.

Lohse, A. P. (2024). Higher Education in an Age of Disruption. Springer Nature Switzerland. https://doi.org/10.1007/978-3-031-57912-7

Mahoney, J., & Thelen, K. A. (Eds.). (2010). Explaining institutional change: Ambiguity, agency, and power. Cambridge Univ. Press.

Maxwell, J. A. (2013). Qualitative research design: An interactive approach (3rd edition). Sage.

McClelland, D. C. (1973). Testing for competence rather than for "intelligence". The American Psychologist, 28(1), 1–14. https://doi.org/10.1037/h0034092

Medizinischer Fakultätentag (MFT) Deutschland. (10.2013). Medizinischer Fakultätentag. Wir bilden Ärzte aus, forschen und sorgen für Ihre Gesundheit: Beilage zur duz – Unabhängige Deutsche Universitätszeitung/Magazin für Forscher und Wissenschaftsmanager. https://medizinische-fakultaeten.de/wp-content/uploads/2018/01/duz-spec_medfaktag_web.pdf

Medizinischer Fakultätentag (MFT) Deutschland. (2024). Tagungsbericht 85. Ordentlicher Medizinischer Fakultätentag am 13. und 14. Juni 2024 in Bonn. https://medizinische-fakultaeten.de/wp-content/uploads/2024/12/0MFT-Tagungsband_final_Unterschrift.pdf

Medizinischer Fakultätentag (MFT) Deutschland; Deutsche Hochschulmedizin e.V. (2023, August 9). Stellungnahme der Deutschen Hochschulmedizin zum Referentenentwurf zur Ärztlichen Approbationsordnung vom 15.6.2023 [Press release]. https://medizinische-fakultaeten.de/wp-content/uploads/2023/08/Stellungnahme-DHM-RefE-AeApprO-230809.pdf

Meikle, J. L. (1995). American plastic: A cultural history. Rutgers University Press.

Meuser, M., & Nagel, U. (2009). The Expert Interview and Changes in Knowledge Production. In A. Bogner, B. Littig, & W. Menz (Eds.), Interviewing Experts (pp. 17–42). Palgrave Macmillan UK. https://doi.org/10.1057/9780230244276_2

Miani, C., Hinrichs, S., Pitchforth, E., Bienkowska-Gibbs, T., Disbeschl, S., Roland, M., & Nolte, E. (2015). Best practice. Medical training from an international perspective. RAND Europe. https://www.rand.org/pubs/research_reports/RR622.html

Miller, G. E. (1990). The assessment of clinical skills/competence/performance. Academic Medicine: Journal of the Association of American Medical Colleges, 65(9), 63–67.

Mol, A. (2014). Language Trails: 'Lekker' and Its Pleasures. Theory, Culture & Society, 31(2-3), 93–119. https://doi.org/10.1177/0263276413499190

Molzberger, K. (2020). Autonomie und Kalkulation. Arbeit und Organisation. Transcript.

Müller, B. (2012). „Wir befinden uns in einer permanenten Reform [...]". Reformen in der Medizinerausbildung zwischen 1989 bis 2009. https://www.bosch-stiftung.de/de/publikation/wir-befinden-uns-einer-permanenten-reform-reformen-der-medizinerausbildung-zwischen

Nettleton, S. (2021). The sociology of health and illness (Fourth edition). Polity.

Nott, J. (2023). Architecture for Anatomy: History, Affect, and the Material Reproduction of the Body in Two Medical School Buildings. Body & Society, 29(2), 99–129. https://doi.org/10.1177/1357034X231161312

Nott, J., & Harris, A. (2023). Teaching the normal and the pathological: educational technologies and the material reproduction of medicine. Science as Culture, 32(2), 214–239. https://doi.org/10.1080/09505431.2023.2171859

Nott, J., Harris, A., Vaden Allison, R., van Lente, H., Sánchez Burmester, C. F., Wojcik, A., & Wyatt, S. (Eds.). (2022). Making sense of

medicine: Material culture and the reproduction of medical knowledge. Intellect.
Oh, H. (2014). Hospital consultations and jurisdiction over patients: Consequences for the medical profession. Sociology of Health & Illness, 36(4), 580–595. https://doi.org/10.1111/1467-9566.12087
Oh Nelson, H. (2022). Conflicted care: Doctors navigating patient welfare, finances, and legal risk. Stanford University Press.
O'Regan, G. (2021). A Brief History of Computing (3rd ed. 2021). Springer International Publishing; Imprint Springer. https://doi.org/10.1007/978-3-030-66599-9
Orren, K., & Skowronek, S. (2018). Beyond the Iconography of Order: Notes for a "New Institutionalism". In L. C. Dodd & C. Jillson (Eds.), The Dynamics of American Politics (pp. 311–330). Routledge. https://doi.org/10.4324/9780429496677-14
Otte, G., Sawert, T., Brüderl, J., Kley, S., Kroneberg, C., & Rohlfing, I. (2023). Gütekriterien in der Soziologie. Zeitschrift Für Soziologie, 52(1), 26–49. https://doi.org/10.1515/zfsoz-2023-2006
Owen, H. (2016). Simulation in Healthcare Education. Springer International Publishing. https://doi.org/10.1007/978-3-319-26577-3
Parsons, T. (1939). The Professions and Social Structure. Social Forces, 17(4), 457–467.
Patton, M. Q. (2015). Qualitative research & evaluation methods: Integrating theory and practice (4. ed.). Sage.
Penter, V., & Augurzky, B. (2014). Gesundheitswesen für Praktiker. Springer Fachmedien Wiesbaden.
Petersohn, S., & Heinze, T [Thomas] (2018). Professionalization of bibliometric research assessment. Insights from the history of the Leiden Centre for Science and Technology Studies (CWTS). Science and Public Policy, 45(4), 565–578. https://doi.org/10.1093/scipol/scx084
Pfadenhauer, M. (2003). Professionalität.: Eine Wissenssoziologische Rekonstruktion Institutionalisierter Kompetenzdarstellungskompetenz. Leske + Budrich.
Pfadenhauer, M. (2009). At Eye Level: The Expert Interview — a Talk between Expert and Quasi-expert. In A. Bogner, B. Littig, & W. Menz

(Eds.), Interviewing Experts (pp. 81–97). Palgrave Macmillan UK. https://doi.org/10.1057/9780230244276_4

Pierson, P. (2000). Increasing Returns, Path Dependence, and the Study of Politics. American Political Science Review, 94(2), 251–267. https://doi.org/10.2307/2586011

Pierson, P. (2015). Power and path dependence. In J. Mahoney & K. Thelen (Eds.), Advances in Comparative-Historical Analysis (pp. 123–146). Cambridge University Press. https://doi.org/10.1017/CBO9781316273104.006

Prentice, R. (2013). Bodies in formation: An ethnography of anatomy and surgery education. Duke University Press.

Renschler, H. E. (1987). Die Praxisphase im Medizinstudium: Die geschichtliche Entwicklung der klinischen Ausbildung mit der Fallmethode. Springer Berlin Heidelberg.

Rotgans, J. I. (2012). The themes, institutions, and people of medical education research 1988–2010: Content analysis of abstracts from six journals. Advances in Health Sciences Education : Theory and Practice, 17(4), 515–527. https://doi.org/10.1007/s10459-011-9328-x

Roth, P. H. (2021). Medicine as science.: The Making of Disciplinary Identity from Scientific Medicine to Biomedicine. Nomos.

Roth, W.-M. (2013). Translation in Qualitative Social Research: The Possible Impossible. Advance online publication. https://doi.org/10.17169/fqs-14.2.1986 (Forum Qualitative Sozialforschung / Forum: Qualitative Social Research, Vol 14, No 2 (2013)).

Sander, T. (2023). Milieuzugehörigkeiten von Pflegekräften und MedizinerInnen: Chancen und Grenzen für Kooperation. In S. Dangendorf & T. Sander (Eds.), Akademisierung der Pflege: Berufliche Identitäten und Professionalisierungspotentiale im Vergleich der Sozial- und Gesundheitsberufe (2. Auflage, pp. 157–174). Juventa Verlag.

Schittenhelm, K. (2025). Die (Nicht-)Übersetzbarkeit wissenschaftlicher Ergebnisse. In C. Bading, K. Kazzazi, & J. Wintzer (Eds.), (Fremd-)Sprache und Qualitative Sozialforschung (pp. 329–343). Springer Berlin Heidelberg. https://doi.org/10.1007/978-3-662-65766-9_18

Schmidt, V. A. (2008). Discursive Institutionalism: The Explanatory Power of Ideas and Discourse. Annual Review of Political Science, 11(1), 303–326. https://doi.org/10.1146/annurev.polisci.11.060606.135342

Schneijderberg, C., Steinhardt, I., & Wieczorek, O. (2022). Qualitative und quantitative Inhaltsanalyse: digital und automatisiert. Beltz Juventa.

Schubert, C. (2006). Die Praxis der Apparatemedizin: Ärzte und Technik im Operationssaal. Campus.

Scott, W. R. (1995). Institutions and organizations. Foundations for organizational science. Sage. http://www.loc.gov/catdir/enhancements/fy0658/95010016-d.html

Scott, W. R., Ruef, M., Mendel, P. J., & Caronna, C. A. (2000). Institutional change and healthcare organizations: From professional dominance to managed care. Univ. of Chicago Press.

Segarra, L. M., Schwedler, A., Weih, M., Schmidt, A., & Hahn, E. G. (2008). Der Einsatz von medizinischen Trainingszentren für die Ausbildung zum Arzt in Deutschland, Österreich und der deutschsprachigen Schweiz. GMS Journal for Medical Education(2).

Shea, K. L., & Rovera, E. J. (2015). Vaginal Examination Simulation Using Citrus Fruit to Simulate Cervical Dilation and Effacement. Cureus, 7(9), 1–9. https://doi.org/10.7759/cureus.314

Simon, M. (2021). Das Gesundheitssystem in Deutschland: Eine Einführung in Struktur und Funktionsweise (7., überarbeitete und erweiterte Auflage). Hogrefe.

Singh, H., Kalani, M., Acosta-Torres, S., El Ahmadieh, T. Y., Loya, J., & Ganju, A. (2013). History of simulation in medicine: From Resusci Annie to the Ann Myers Medical Center. Neurosurgery, 73 Suppl 1, 9–14. https://doi.org/10.1227/NEU.0000000000000093

Spencer, J., & Dales, J. (2006). Meeting the needs of simulated patients and caring for the person behind them? Medical Education, 40(1), 3–5. https://doi.org/10.1111/j.1365-2929.2005.02375.x

St. Pierre, M. (2018). Blick zurück: Die Geschichte der Patientensimulation. In M. St.Pierre & G. Breuer (Eds.), Simulation in der Medizin (pp. 1–17). Springer.

St. Pierre, M., & Breuer, G. (Eds.). (2018). Simulation in der Medizin. Springer.

Stark, A. (2018). New institutionalism, critical junctures and post-crisis policy reform. Australian Journal of Political Science, 53(1), 24–39. https://doi.org/10.1080/10361146.2017.1409335

Starystach, S., & Dettmer, S. (2023). Translational aus der Krise und hin zu einem zukunftsfähigen Gesundheitssystem? Der schwierige Umbau der deutschen Universitätsmedizin. In I. Bode, R. Jungmann, & K. Serrano Velarde (Eds.), Staatliche Organisationen und Krisenmanagement (pp. 135–156). Nomos Verlagsgesellschaft mbH & Co. KG. https://doi.org/10.5771/9783748936794-135

Stehr, N., & Grundmann, R. (2010). Expertenwissen: Die Kultur und die Macht von Experten, Beratern und Ratgebern. Velbrück Wissenschaft.

Stein, R. (2009). Reform des Medizinstudiums in Berlin – Eine unendliche Geschichte. Berliner Ärzte. Die Offizielle Zeitschrift Der Ärztekammer Berlin(11), 15–18.

Stollberg, G. (2001). Medizinsoziologie. Einsichten Themen der Soziologie. Transcript.

Streeck, W., & Thelen, K. (2005). Beyond continuity: Institutional change in advanced political economies. Oxford University Press.

Symonds, J. E., & Gorard, S. (2010). Death of mixed methods? Or the rebirth of research as a craft. Evaluation & Research in Education, 23(2), 121–136. https://doi.org/10.1080/09500790.2010.483514

Tashakkori, A., & Teddlie, C. (2009). Integrating Qualitative and Quantitative Approaches to Research. In L. Bickman & D. J. Rog (Eds.), The SAGE Handbook of Applied Social Research Methods. (2. Ed., pp. 283–317). Sage.

Temkin, O. (1949). Medicine and the problem of moral responsibility. Bulletin of the History of Medicine, 23(1), 1–20.

Thelen, K. (1999). Historical Institutionalism in Comparative Politics. Annual Review of Political Science, 2(1), 369–404. https://doi.org/10.1146/annurev.polisci.2.1.369

Thieme via medici. (2024). IMPP-Fragen – das ändert sich ab 2025 [Press release]. https://viamedici-testen.thieme.de/impp-fragen/

Thistlethwaite, J., & Ridgway, G. (2006). Making it real: A practical guide to experiential learning of communication skills. Radcliffe Pub.

Timmermans, S., & Berg, M. (2003a). The gold standard: The challenge of evidence-based medicine and standardization in health care. Temple University Press.

Timmermans, S., & Berg, M. (2003b). The practice of medical technology. Sociology of Health & Illness, 25, 97–114. https://doi.org/10.1111/1467-9566.00342

Underman, K. (2022). Gloves: The Context of Touch: Gloves and the Pelvic Exam. In J. Nott, A. Harris, R. Vaden Allison, H. van Lente, C. F. Sánchez Burmester, A. Wojcik, & S. Wyatt (Eds.), Making sense of medicine: Material culture and the reproduction of medical knowledge (pp. 187–193). Intellect.

van Drie, M. (2013). Training the Auscultative Ear. The Senses and Society, 8(2), 165–191. https://doi.org/10.2752/174589313X13589681981019

Weinert, F. E. (2001). Concept of competence: A conceptual clarification. In D. S. Rychen & L. H. Salganik (Eds.), Defining and selecting key competencies (pp. 45–65). Hogrefe and Huber Publishers.

Weingast, B. (2002). Rational Choice Institutionalism. In I. Katznelson & H. V. Milner (Eds.), Political science: The state of the discipline (pp. 660–692). Norton.

Wernitz, M. H., Reinhold, T., & Sydow, H. (2022). Gesundheitsökonomie und das deutsche Gesundheitswesen: Ein praxisorientiertes Lehrbuch für Studium und Beruf (3., erweiterte und aktualisierte Auflage). Verlag W. Kohlhammer.

Wier, G. S., Tree, R., & Nusr, R. (2017). Training Effectiveness of a Wide Area Virtual Environment in Medical Simulation. Simulation in Healthcare : Journal of the Society for Simulation in Healthcare, 12(1), 28–40. https://doi.org/10.1097/SIH.0000000000000207

Wilkesmann, M. (2009). Wissenstransfer im Krankenhaus: Institutionelle und strukturelle Voraussetzungen (1. Aufl.). VS-Research. VS Verlag für Sozialwissenschaften.

Wilkesmann, M., Schiek, D., & Steden, S. (2019). Der Umgang mit Nichtwissen in Worten. In M. Wilkesmann & S. Steden (Eds.), Nichtwissen stört mich (nicht) (pp. 163–216). Springer Fachmedien Wiesbaden. https://doi.org/10.1007/978-3-658-22009-9_8

Wissenschaftsrat. (2018). Neustrukturierung des Medizinstudiums und Änderung der Approbationsordnung. Empfehlungen der Expertenkommission zum Masterplan Medizinstudium 2020. https://www.wissenschaftsrat.de/download/archiv/7271-18.html

Wojcik, A. (2022). Simulations in health professions education. In J. Nott, A. Harris, R. Vaden Allison, H. van Lente, C. F. Sánchez Burmester, A. Wojcik, & S. Wyatt (Eds.), Making sense of medicine: Material culture and the reproduction of medical knowledge (pp. 360–364). Intellect.

World Health Organization. Regional Office for Europe & European Observatory on Health Systems and Policies. (2000). Health care systems in transition: Germany. World Health Organization Regional Office for Europe. https://iris.who.int/handle/10665/108301

Zorn, D., Dobbin, F., Dierkes, J., & Kwok, M.-S. (2006). The New Firm. Power and Sense-making in the Construction of Shareholder Value. Nordiske Organisationsstudier, 3.